Lecture Notes in Artificial Intellig

T0237900

Edited by R. Goebel, J. Siekmann, and W. Wah.

Subseries of Lecture Notes in Computer Science

David Riaño (Ed.)

Knowledge Management for Health Care Procedures

ECAI 2008 Workshop, K4HelP 2008
Patras, Greece, July 21, 2008
Revised Selected Papers

 Springer

Series Editors

Randy Goebel, University of Alberta, Edmonton, Canada
Jörg Siekmann, University of Saarland, Saarbrücken, Germany
Wolfgang Wahlster, DFKI and University of Saarland, Saarbrücken, Germany

Volume Editor

David Riaño
Universitat Rovira i Virgili
Dept. Enginyeria Informática i Matemátiques - ETSE
Av. Països Catalans 26, 43007 Tarragona, Spain
E-mail: david.riano@urv.net

Library of Congress Control Number: 2009930846

CR Subject Classification (1998): I.2, J.3, H.2.8, H.3, H.4, H.5

LNCS Sublibrary: SL 7 – Artificial Intelligence

ISSN   0302-9743
ISBN-10   3-642-03261-3 Springer Berlin Heidelberg New York
ISBN-13   978-3-642-03261-5 Springer Berlin Heidelberg New York

springer.com

© Springer-Verlag Berlin Heidelberg 2009
Printed in Germany

Typesetting: Camera-ready by author, data conversion by Scientific Publishing Services, Chennai, India
Printed on acid-free paper   SPIN: 12697250   06/3180   5 4 3 2 1 0

Preface

The intersection between knowledge management, computer science, and health care defines a technological area of great interest that has not been operated properly. Within this area medical procedures on preventive, diagnostic, therapeutic, or prognostic tasks in health care play an outstanding role. The management of this type of knowledge at the point of care includes four technological scopes, at least. The first one establishes the languages and structures to represent health care procedural knowledge and the integration of these structures with medical information systems. The second consists of the development of algorithms and computer science technologies for the operation of this knowledge. The third scope is concerned with the development of methodologies to maximize the benefit of these algorithms and methodologies. The fourth concerns the integration of the previous algorithms, technologies, and methodologies in computer science systems that allow the application of this knowledge at the point of need, harnessing health care of greater quality and efficiency.

Under this vision, the workshop Knowledge for Management Health Care Processes (K4HelP 2008) was organized as part of the 18th European Conference on Artificial Intelligence (AIME 2008) and centered on the following topics: formalization of medical processes and health care knowledge-based models; knowledge representation and ontologies for health care processes; health care knowledge representation standards; time-based health care knowledge representation and exploitation; digital libraries and repositories on health care procedural data, knowledge, and services; knowledge combination and adaptation for health care processes; health care procedural knowledge extraction from textual documents; procedural knowledge extraction from health care database; application of procedural knowledge in health care; and procedural knowledge for medical e-training and clinical practice.

The workshop was the fourth in a series of workshops and publications devoted to the formalization, organization, and deployment of procedural knowledge in health care. Previous workshops and publications have been the IEEE CBMS 2007 special track on Machine Learning and Management of Health Care Procedural Knowledge, the AIME 2007 workshop entitled From Medical Knowledge to Global Health Care, and Springer's LNAI 4924.

The K4HelP 2008 workshop was chaired by David Riaño, and it received 14 papers, among which 10 were selected according to their relevance, quality, and originality. This volume contains extended versions of these accepted papers, plus two invited papers that contribute to providing a broader vision of the above-mentioned aspects that are relevant to the progress of knowledge management for health care procedures.

In this volume, the papers are structured in three sections: technologies to manage health care procedural knowledge, methodologies to manage health care procedural knowledge, and computer systems to manage health care procedural knowledge; each one of them containing four papers.

I want to thank everyone who contributed to the K4HelP 2008 workshop: the authors of the submitted papers, the invited authors, the members of the Organizing Committee, the members of the Program Committee, and the sponsoring institutions.

April 2009   David Riaño

Organization

The workshop Knowledge Management for Health Care Processes (K4HelP 2008) was organized by David Riaño from the Department of Computer Science and Mathematics, Rovira i Virgili University.

Organizing Committee

Conference Chair: David Riaño (Rovira i Virgili University, Spain)
Support team: John A. Bohada (Rovira i Virgili University, Spain)
    Francis Real (Rovira i Virgili University, Spain)
    Aida Kamisalic (Rovira i Virgili University, Spain)

Program Committee

Syed Sibte Raza Abidi	Dalhousie University, Canada
Ameen Abu-Hanna	University of Amsterdam, The Netherlands
Roberta Annicchiarico	Santa Lucia Hospital, Italy
Fabio Campana	CAD RMB, Italy
Karina Gibert	Technical University of Catalonia, Spain
Femida Gwadry-Sridhar	University of Western Ontario, Canada
Lenka Lhotska	Czech Technical University, Czech Republic
Patrizia Meccoci	University of Perugia, Italy
Antonio Moreno	Rovira i Virgili University, Spain
David Riaño	Rovira i Virgili University, Spain
Maria Taboada	University of Santiago de Compostela, Spain
Aida Valls	Rovira i Virgili University, Spain
Laszlo Varga	MTA STAKI, Hungary

Sponsoring Institutions

Research Group on Artificial Intelligence, Banzai, Tarragona, Spain
Rovira i Virgili University, Tarragona, Spain
FP6 IST K4CARE Project
Diputació de Tarragona, Tarragona, Spain

Table of Contents

Technologies to Manage Health Care Procedural Knowledge

Methodologies to Manage Health Care Procedural Knowledge

Computer Systems to Manage Health Care Procedural Knowledge

Operationalizing Prostate Cancer Clinical Pathways: An Ontological Model to Computerize, Merge and Execute Institution-Specific Clinical Pathways

Samina Raza Abidi[1], Syed Sibte Raza Abidi[1], Lorna Butler[2], and Sajjad Hussain[1]

[1] NICHE Research Group, Faculty of Computer Science, Dalhousie University, Halifax, Canada
[2] College of Nursing, University of Saskatchewan, Saskatoon, Canada

Abstract. The computerization of paper-based Clinical Pathways (CP) can allow them to be operationalized as a decision-support and care planning tool at the point-of-care. We applied a knowledge management approach to computerize the prostate cancer CP for three different locations. We present a new prostate cancer CP ontology that features the novel merging of multiple CP based on the similarities of their diagnostic-treatment concepts, whilst maintaining the unique aspects of each specific CP, to realize a common unified CP model. In this paper we will highlight the main components of our prostate cancer CP ontology, and discuss the concept of CP branching and merging nodes. We conclude that our computerized CP can be executed through a logic-based engine to realize a point-of-care decision-support system for managing prostate cancer care.

1 Introduction

Prostate cancer is the most common type of cancer among Canadian men, with an estimated 22,300 newly diagnosed cases and 4,300 deaths in Canada in 2007 alone. In the Canadian system, the diagnosis and treatment of prostate cancer follows an integrated approach involving multiple disciplines dispersed across multiple care setting and engaging multiple health professionals with different specialities and roles. This integrated approach demands an effective partnership between various disciplines such as family medicine, urology, radiation oncology, nursing, and psychological support resources. Despite the clinical significance of such an integrated approach, its on-the-ground implementation presents various challenges, such as (a) how to navigate and manage a patient's care activities throughout the longitudinal care trajectory? and (b) how to coordinate the respective activities of the different care providers in a timely and efficient manner?

In an attempt to support the coordination and integration of healthcare services spanning multidisciplinary settings and care providers, healthcare institutions develop Clinical Pathways (CP) as a means to both chart and streamline the diagnostic-treatment cycle. CP are evidence-based patient care algorithms/charts that describe the care process for specific medical conditions within a localized setting [1]. At present, most

D. Riaño (Ed.): K4HelP 2008, LNAI 5626, pp. 1–12, 2009.

CP are paper-based and therefore cannot be conveniently shared and directly deployed at the point-of-care, regardless of the location of the patient and the attending care provider. We argue that the computerization of paper-based CP can help to operationalize them as (a) point-of-care clinical guide; (b) patient information sharing medium between different care providers; (c) patient navigation and care coordination tool; and (d) a decision-support tool to help provide standardized, timely, cost-effective and safe clinical care to prostate cancer patients [2].

In this paper we present our knowledge modeling work leading to the development of a prostate cancer care planning and management system. The overall project involves three phases: (i) the development of prostate cancer CP for three different Canadian cancer care institutions in Halifax, Winnipeg and Calgary. In this knowledge engineering phase, oncologists, urologists and nursing experts were engaged to elicit the CP in their respective institutions, thus yielding three location-specific prostate cancer CP; (ii) the modeling of the CP knowledge in order to computerize and subsequently execute the CP (with patient data) at the point-of-care. We present our ontology-based knowledge modeling approach that led to the development of a comprehensive OWL-based prostate cancer care ontology. The feature of our modelling approach is that it allows the merging of these location-specific CP along common processes, actions and recommendations; and (iii) the execution of the ontologically-modeled CP using a logic-based execution engine that connects with a patient-data source to guide both the respective care-provider and the patient through the prostate cancer care pathway.

In this paper we will describe our ontology based CP knowledge modeling approach. We will highlight the main components of our ontology, especially the unique merging and branching nodes that are used to merge three location-specific CP into a single model.

2   Prostate Cancer Clinical Pathways

In this project we developed prostate cancer CP that illustrate activities concerning the diagnosis, management and follow up of prostate cancer patients at three different locations–i.e. Halifax, Calgary and Winnipeg regional health setting. Each location-specific CP characterizes the following: (a) Organizational level processes to be enacted by a team of multidisciplinary actors; and (b) Patient management processes that require a specialized care team member to perform a specific action on the patient. A systematic organization of this information yielded a prostate cancer CP as a flow-chart that contains four well-known components–namely *actions, decisions, branching/merging nodes* and *recommendations/plans* (see figure 1).

All three location-specific CP were divided into four consultations–namely (1) visit to family physician, (2) visit to primary urologist, (3) visit to secondary urologist and (4) treatment option. In each consultation a set of tasks were performed by an identified team member(s) to achieve a defined outcome. Each CP begins with a consultation by a family physician and concludes with a consultation by an urologist to determine treatment options and follow-up routines. For each consultation, the CP records the stipulated clinical practices and care resources in terms of the sequencing, decision criterion, time intervals, actors, expected outcome and recommendations associated with specific

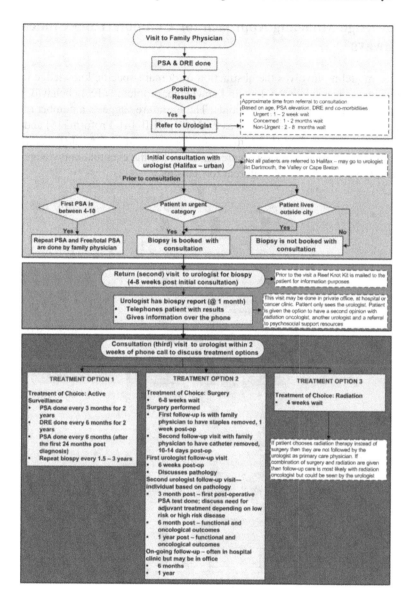

Fig. 1. Prostate Cancer Clinical Pathway for Halifax

care tasks that need to be performed during a consultation. It was interesting to note that despite certain variations, the three CP exhibited a good deal of overlap at the task-level, thus vindicating that these CP conform to widely accepted Canadian practices for prostate care. The overlap between the CP allowed us to pursue the **merging** of the different CP to create a common unified location-independent CP that distinguishes between a set of tasks common to all locations, whilst allowing location-specific branches to model those tasks that are unique to a particular location.

3   Knowledge Modeling Approach for Computerizing Clinical Pathways

Knowledge modeling involves the abstraction of domain-specific knowledge in terms of concepts that encapsulate the domain knowledge, problem-solving behavior, operational processes, and functional constraints. The literature suggests a number of health knowledge modeling formalisms, such as EON [3], GLIF [4], Proforma [5] and SAGE [6], each using a specific knowledge modeling approach. Most formalisms refer to the use of ontologies [7] for knowledge representation, but the eventual knowledge execution capability vary between the different formalisms.

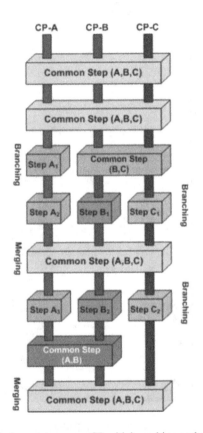

Fig. 2. A unified prostate cancer CP with branching and merging nodes

We followed a knowledge management approach, in particular the use of ontologies for knowledge modeling. Our prostate cancer CP ontology uses well-understood health knowledge constructs to capture the diagnostic, treatment and operational concepts, and relates these concepts using semantic and pragmatic relationships. We have developed unique knowledge constructs that are not only executable but also allow the merging of CP based on attributes such as institution's location and patient co-morbidities.

We used a middle-out approach for ontology engineering [8], whereby the knowledge model is iteratively developed-i.e. starting from generic to specific concepts and relationships–using the three prostate cancer CP. Instead of modeling each CP as a unique model, our CP modeling approach focused on ways to **merge** the three different location-specific CP to realize a unified ontological model for prostate cancer CP. The rationale for merging CP is to create a flexible knowledge model that not only responds to the clinical or administrative events in the care process, but also factors the various constraints, such as the location of the patient, when discharging recommendations/actions. Technically speaking, the ability to merge the CP in a flexible yet semantically and pragmatically correct knowledge model was the main knowledge modeling challenge [9]. CP merging was pursued by modeling the task-level similarities between the three CP as a single common ontology, however whenever we encountered a point when a location-specific CP was pursing a unique set of tasks we created a branch ontology to model the unique task, treatment or follow-up options practiced at a specific location. A branch ontology proceeds along a location-specific path until it reaches a **merging node**–i.e. a task or a plan that is common to all locations–that allows multiple branches to merge to once again realize a common path modeled by the overall *common ontology*. In this way, we developed a novel CP modeling approach that allowed multiple CP from the same domain to be jointly modeled whilst maintaining the unique behaviors of independent CP. Figure 2 shows a schematic of a unified CP for three different sites (A, B, C), highlighting both branching and merging nodes.

4   A Prostate Cancer CP Ontology

We used the Web Ontology Language (OWL) via the ontology editor Protégé to develop our prostate cancer CP ontology. Below we present the salient aspects of our ontology. Class names are denoted using UPPERCASE, relationships with *Italics* and individuals within 'quotation marks'.

4.1   Descriptions of the Classes and Their Individuals

Our ontology begins with class PLAN which corresponds to all four consultations with a team of multidisciplinary CLINICIANS. DECISION-CRITERIA models the choices to be made in order to reach the next step, for instance the individuals 'between 4 and 10' and 'greater than 10' are used as decision criteria for a decision-node 'PSA/FreeTotalPSA' which is an individual of INVESTIGATION. The evaluation of DECISION-CRITERIA results in either a TASK to be performed or a TEST-RESULT to be generated. TASK represents the different care tasks performed by the care team. TASK is further classified as CONSULTATION-TASK, NON-CONSULTATION-TASK, REFERRAL-TASK and FOLLOW-UP-TASK. The class FOLLOW-UP represents follow-up visits after each treatment option, e.g. 'FirstPostSurgeryFollowUp'. To control the execution of the pathway, we have defined a class TERMINATION-TASK as a sub-class of TASK, which has two individuals 'PathwayEnds' which specifies the end of the CP and 'TaskEnds' which represents the end of a task. A PLACE is further categorized into CARESETTING with exemplar individual being 'RapidAccessClinic', and PATHWAY-REGION with exemplar individuals 'Calgary', 'Halifax' and 'Winnipeg'.

PATIENT-CONDITION-SEVERITY specifies the condition of the patient as being 'Urgent', 'Concerned' and 'NonUrgent'. TREATMENT represents treatment options, for instance 'ActiveSurveillance', 'Brachytherapy' etc. INVESTIGATION captures diagnostic tests, e.g. 'Biopsy', 'PSA/
FreeTotalPSA'.

4.2 Modeling of Temporal Concepts in the CP

The temporal concepts in the CP are represented by three classes:

1. INTERVAL-EVENT which defines an interval between activities or wait before a particular task, as a named event, e.g. wait interval for surgery.
2. INTERVAL-DURATION which defines the duration of an interval event, e.g. six to eight week which is wait time for surgery. Another temporal constraint inherent in a CP is the frequency of activities within a task.
3. FREQUENCY depicts the frequency of the follow-up activities noted in the prostate cancer CP, for instance to represent the concept EveryThreeMonths. Preserving FREQUENCY as a separate class ensures that future changes or addition to frequency of an activity can be easily incorporated in the model.

4.3 Description of the Relationships between the Classes

Our prostate cancer CP ontology models a large number of relationships between classes; here we present some salient relationships. PLAN, TEST-RESULT and PATIENT-CONDITION-SEVERITY have relation *isFollowedByTask* with TASK, because an individual of any of these classes is followed by a TASK. For example, if PATIENT-CONDITION-SEVERITY is 'NonUrgent' then it *isFollowedByTask* 'BiopsyIsNotBookedWithSecondConsultation' which is an individual of CONSULTATION-TASK. A task can be followed by another task, therefore TASK has the relation *isFollowedByTask* with itself also. TASK, TREATMENT and FOLLOW-UP have relationship *hasInterval* with INTERVAL-EVENT, e.g. 'ReferToUrologist' as an individual of REFERRAL-TASK with *hasInterval* to represent 'TimeToReferToUrologist' which

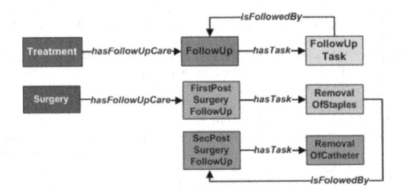

Fig. 3. Interrelationships between the classes TREATMENT, FOLLOW-UP and TASK

is an individual of INTERVAL-EVENT. TREATMENT is related to FOLLOW-UP via *hasfollowUpCare*. A follow-up might refer to follow-up task(s), therefore FOLLOW-UP is related to TASK via *hasTask*. For example 'FirstPostSurgeryFollowUp', which is an individual of FOLLOW-UP *hasTask* 'RemovalOfStaples' which is an individual of FOLLOWUP-TASK. TASK and FOLLOW-UP have relationship *isFollowedBy* with FOLLOW-UP as its range. For example the TASK 'RemovalOfStaples' *isFollowedBy* 'SecondPostSurgeryfollowUp' which is an individual of FOLLOW-UP. In turn, 'SecondPostSurgeryFollowUp' *hasTask* 'RemovalOfCatheter' which *isFollowedBy* 'ThirdPostSurgeryFollowUp' which is an individual of FOLLOW-UP. A snapshot of this scenario is shown in figure 3.

5   Modelling Branching and Merging within the CP Ontology

We have developed a single prostate cancer CP ontology that is able to uniquely model the independent characteristics of all the three different CP. Our modeling approach allows the merging of the three location-specific CP into a unified CP ontology based on the commonality of their inherent concepts at the level of clinical pragmatics. Yet, in order to model the non-overalpping concepts between the CP we have introduced a **branching** function/node that allows an independent CP to pursue tasks specific to it. And, through a **merging** function/node we allow the branched CP to once again merge with other concurrent CP to realize a high-level unified CP ontology. Figure 2 earlier presented the concept of CP merging and branching.

5.1   Branching Based on Decision Criteria

In our CP ontology certain individuals of classes INVESTIGATION, TASK and FOLLOW-UP can also be regarded as **decision nodes** in a CP, therefore these classes are related to class DECISION-CRITERIA through relationship *hasDecisionCriteria* (as illustrated in figure 4). As mentioned earlier, the class DECISION-CRITERIA models the available choices (or paths) when determining the next step–one of the given choices is selected (based on user input) in order to proceed to the next specified step. We explain this concept through an example illustrating how next step choices are handled in our ontology. Consider 'TakePatientConsent' (an individual of CONSULTATION-TASK) as a decision node in the CP, with two possible choices–i.e. 'PatientGivesConsent' and 'PatientDoesNotGiveConsent' (individuals of DECISION-CRITERIA) as the set of potential values for the relation *hasDecisionCriteria*. During execution, when we arrive at the above-mentioned decision node we need to select one of these choices in order to direct the flow of the CP in a particular direction, which is modeled by TASK through property *hasAction*–note that DECISION-CRITERIA is related to TASK through property *hasAction*. Suppose, in response to the value 'TakePatientConsent' the relation *hasDecisionCriteria* gets the value 'PatientGivesConsent', then the value for the next*hasAction* relation will be 'BookBiopsyWithSecondConsultation', on the other hand if value for *hasDecisionCriteria* is 'PatientDoesNotGiveConsent', then the value for *hasAction* will be 'DoNotBookBiopsyWithSecondConsutation'. In this way we are able to model branching effects within a CP based on decision nodes.

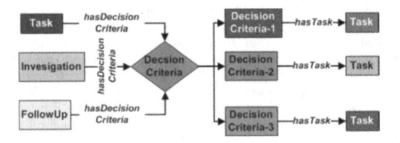

Fig. 4. Modeling of decision criteria

5.2 Branching Based on Location

Another type of branching involves a CP diverging from the unified ontology based on the location of the patient for a given task, treatment or follow-up options. We model this behavior through **branching nodes** that denote an intersection between two classes to represent a unique individual that is the function of two intersected classes. We have developed three unique classes–i.e. REGION-TASK-INTERSECTION, REGION-TREATMENT-INTERSECTION and REGION-FOLLOWUP-INTERSECTION that serve as branching nodes based on location. The REGION-TASK-INTERSECTION represents an intersection between REGION and TASK to signify a unique individual, such as a unique TaskA that is perfomed at RegionB. Likewise, REGION-TREATMENT-INTERSECTION will have an individual that is a unique TreatmentX that is offered in a specific region. Note that if TreatmentX was common for all three regions then there was no need to use an intersection to denote a branch, rather TreatmentX would have been part of the unified CP. The branching nodes have relations *hasLocation, hasTask, includeTreatmentOptions* and *hasFollowUpCare*. REGION-TASK-INTERSECTION has an object property *isFollowedByConsultation*, the range of which is class PLAN, to represent the possibility that a task at a particular location can be followed by a new consultation as opposed to a task. These relationships were carefully determined to ensure that we always have unique individuals of the classes based on the combination of the values of these relationships. For example, an individual of any of the classes PLAN, TASK, PATIENT-CONDITION-SEVERITY and TEST-RESULT can be followed by a task that is specific to a certain location only, thus initiating the branching of that particular segment of the location-specific CP from the unified CP.

We explain the concept of branching using intersections through the following example, also depicted in figure 5. In the three CP, it is noted that the activities following consultation-2 are different, such that the tasks in Calgary are different from the ones in Winnipeg and Halifax. So during CP execution, when a patient enters 'Consultation-2' which is an individual of PLAN, the next task in this plan depends on the location of the patient. This is modeled by PLAN having a relation *isFollowedByRegionTaskIntersection* which in this case has values 'RegionTask Intersection-1' and 'RegionTaskIntersection-3', both of which are individuals of the branching node REGION-TASK-INTERSECTION. At this point, the unified CP is

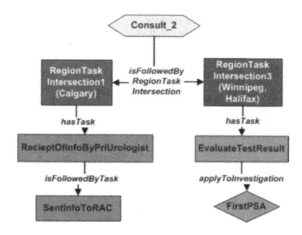

Fig. 5. Branching of CP at the level of Consultation-2

divided into two branches—one branch for Calgary and the other one for Winnipeg and Halifax. The first branch is modeled by the individual 'RegionTaskIntersection-1' (of REGION-TASK-INTERSECTION) that has 'Calgary' as the value for *hasLocation*, and the unique task is 'RecieptOfInformationByPriUrologist' as the value for *hasTask* relation. The second branch is modeled by the individual 'RegionTaskIntersection-3' that has 'Halifax' and 'Winnipeg' as the value for *hasLocation*, and it has 'EvaluateTestResult' as the value for the *hasTask* relation. In this way, we were able to represent the unique activities at a specific location whilst maintaining a common CP structure representing the overlapping activities. It may be noted that these two branches may subsequently merge during the CP execution to realize a unified CP model (see figure 2).

5.3   Merging of the Different CP Branches

The merging of different CP is possible at the level of common tasks or plans. As stated earlier, if a CP branches off then a merging node allows it to merge back with the unified CP if (a) no further activities are left in the branch; or (b) the next activity is a common task or consultation. In figure 6, we illustrate an example of a merging node, whereby during 'Consultation-3' after the task 'RecieptOfBiopsyReportByUrologist' the CP ontology models three separate location-based branches because at each location the following task is different. All the three branches are individuals of REGION-TASK-INTERSECTION, namely 'RegionTaskIntersection-9' that *hasLocation* 'Halifax', 'RegionTaskIntersection-10' with *hasLocation* as 'Calgary', and 'RegionTask Intersection-12' with *hasLocation* having 'Winnipeg' as the individual value. These branches have unique individuals for *hasTask* and *isFollowedByConsultation* relations. However, as shown in figure 6, later on these branches converge on 'Consult-4' (an individual of PLAN) which serves as a merging node to once again realize a unified CP. Note that in Calgary the task 'RecieptOfBiopsyReportByUrologist' is followed directly by 'Consult-4', while in Winnipeg the task before the merge is 'EvaluateBiopsyReport'.

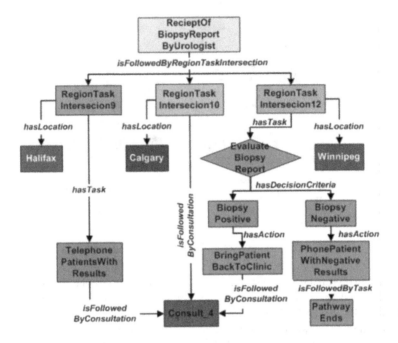

Fig. 6. Merging of three branches at Consultation-4

This task is a decision node where the pathway branches again depending on the result of the biopsy report; if the result is positive then the branch will converge at 'Consult-4'.

6   Modeling Other CP Intersections

Our CP ontology accounts for the eventuality that there might be additional location related CP variations concerning the team member performing a task, time interval between the tasks and frequency of an activity within a task. We have modeled such potential CP variations noted when the classes CLINICIAN, INTERVAL-DURATION and FREQUENCY intersect with location. The resulting intersections are REGION-CLINICIAN-INTERSECTION, REGION-INTERVAL-INTERSECTION, REGION-FREQUENCY-INTERSECTION. To REGION-CLINI CIAN-INTERSECTION accounts for the possibility that a specific TASK, TREAT-MENT or FOLLOW-UP can be performed by a different clinician at a specific region. Our CP ontology relates these classes to REGION-CLINICIAN-INTERSECTION with an object property *hasRegionClinicianIntersection*. Each individual of this class is guaranteed to be unique through the relations *isPerformedBy* and *applyToClinicalSetting* which have ranges CLINICIAN and CARE-SETTING, in addition to the relation *hasLocation*. These properties allow expressing different combinations of location, clinicians and care-setting as unique individuals of class REGION-CLINICAN-INTER SECTION. An individual of TASK, TREATMENT or FOLLOW-UP can then have a unique relationship in terms of location of the patient, a certain type of clinician who is going to perform a task, in a particular care-setting. Similarly an individual of INTERVAL-EVENT through the relation *hasRegionIntervalIntersection* can have a particular interval duration depending on the location of the patient. It may be noted that we do not regard these intersections as branching nodes in a strict sense because they do not lead to separate CP branches.

7   Concluding Remarks

We presented a prostate cancer CP ontology that allows the computerization and execution of a location-specific CP for a single disease. We presented a novel CP modeling approach that first established commonalities in the care processes across the three different locations and then attempted to merge these individual location-specific CP, at the medical and pragmatic levels, to realize a unified high-level prostate cancer CP ontology. In this work we targeted prostate cancer, but the approach is generic and can be applied to any other medical domain. We argue that our unified CP ontology allows to (a) improve the sustainability of the knowledge model in the face of future additions and updates by maintaining a central and common core of care activities; (b) identify the most salient care activities across different institutions, and then suggest them as 'widely practiced step' to other locations; (c) identify the specialized tasks at each location; and (d) compare, and even measure, the outcomes, of the care practices at different locations. Such a performance audit can now be done in a more focused manner by introspecting the location-specific branches within the CP ontology, and this can help understand what are the steps that certain regions are doing differently and what is the impact of their actions compared to other institutions.

From a modeling standpoint, we introduced the concept of branching and merging nodes which were modeled as *inter-class intersections*. For instance, we identified that the CP tend to branch out at the Task, Treatment and FollowUp aspects subject to the location of the patient, and that the different location-specific CP branches merge at the Task and Plan aspects of a CP. The CP ontology was evaluated for semantic correctness and completeness by instantiating the three prostate cancer CP, and was deemed to be representational adequate and accurate. These computerized CP can potentially be executed, when connected with patient data, through a logic-based reasoning engine or a workflow engine. Our future work involves the development of a CP execution engine, based on the CP ontology, and a web-based prostate cancer management system aimed to streamline prostate cancer care at the three Canadian institutions.

Acknowledgement. We thank the Communications in Cancer Socio-Behaviorial Research Team for their partial funding of this research.

References

1. de Bleser, L., Depreitere, R., de Waele, K., Vanhaecht, K., Vlayen, J., Sermeus, W.: Defining pathways. Journal of Nursing Management 14, 553–563 (2006)
2. Chu, S.: Reconceptualising clincal pathway system design. Collegian 8, 33–36 (2001)

3. Tu, S., Musen, M.: Modeling data and knowledge in the eon guideline architecture. In: Medinfo 2001 (2001)
4. Boxwala, A.A., Peleg, M., Tu, S., Ogunyemi, O., Zeng, Q., Wang, D., Patel, V.L., Greenes, R.A., Shortliffe, E.H.: Glif3: A representation format for sharable computer-interpretable clinical practice guidelines. Journal of Biomedical Informatics 37, 147–161 (2004)
5. Sutton, D., Fox, J.: The syntax and semantics of the proforma guideline modeling language. Journal of American Medical Informatics Association 10, 433–443 (2003)
6. Tu, S.W., Campbell, J., Musen, M.A.: The structure of guideline recommendations: A synthesis. In: Proc. AMIA Symposium (2003)
7. Wang, H., Fang, Y., Sun, J., Zang, H., Pan, J.: A semantic web approach to feature modeling and verification. In: Workshop on Semantic Web Enabled Software Engineering (2005)
8. Uschold, M., Gruninger, M.: Ontologies: Principles, methods and applications. Knowledge Engineering Review 11, 93–155 (1996)
9. Sascha, M., Stefan, J.: Process-oriented knowledge support in a clinical research setting. In: Proceedings of 12th IEEE Symposium on Computer-Based Medical Systems. IEEE Press, Los Alamitos (2007)

An Autonomous Algorithm for Generating and Merging Clinical Algorithms

Francis Real and David Riaño

Banzai Research Group
Department of Computer Science and Mathematics,
Universitat Rovira i Virgili,
Av. Països Catalans 26, 43007 Tarragona, Spain
{francis.real,david.riano}@urv.net

Abstract. Procedural knowledge in medicine uses to come expressed as isolated sentences in Clinical Practice Guidelines (CPG) that describe how to act in front of specific health-care situations. Although CPGs gather all evidence available on concrete medical problems, their direct application has been proved to have some limitations. One of these limitations occurs when they have to be applied on co-morbid patients suffering from several simultaneous and mutually related diseases. In such cases, health-care professionals have to follow the indications of multiple CPGs and solve their interactions as they appear in the treatment of concrete patients. Clinical Algorithms (CA) are schematic models of the procedures appearing in a CPG. They are used to organize and summarize the recommendations contained in CPGs. Here, we extend a knowledge-based algorithm to merge CAs with a machine learning procedure to relax the knowledge dependence of that algorithm. The resulting algorithm has been tested on health-care data provided by the SAGESSA Group on hypertension patients. The results obtained prove that it is a good approach to the generation of CA from data though several improvements at the levels of prediction and medical interpretation are possible. Furthermore, the learned knowledge from the data generation process can be reused to improve the results of the merging process for similar diseases.

1 Introduction

In health-care, procedural knowledge exists in multiple tasks as in the process of diagnosing a disease or in the application of a long-term therapy. All the evidence generated on this sort of knowledge for a particular disease is reported on a Clinical Practice Guideline (CPG). CPGs are defined as systematically developed statements to assist practitioners and patient decisions about appropriate health care for specific circumstances [1]. Surprisingly in this definition, the facts that CPGs are systematically developed and that CPGs are for specific circumstances may contradict their final purpose, which is to assist professional decisions. So, on the one hand systematic development in medicine is evidence-based, therefore, when there is not evidence on how to act under certain circumstances, the CPG has a knowledge gap that the physician has to fill with personal experience or consulting other colleagues whenever a patient under such circumstances arrives.

D. Riaño (Ed.): K4HelP 2008, LNAI 5626, pp. 13–24, 2009.

On the other hand CPGs use to be specific for one disease, which is called primary disease, and it may contain indications on how to act if the patient has other diseases, which are considered secondary in the CPG. However, nowadays, the most frequent patient is one with several important diseases that require simultaneous attention. That is to say, patients that require the simultaneous application of several CPGs. Whereas covering knowledge gaps in a CPG is a purely medical task, the simultaneous application of several guidelines affecting a patient is a task that can be systematized with computer intelligence techniques that merge CPGs [2,3]. All these techniques are sustained on the representation of the procedural knowledge of the CPGs with formal languages that computers are able to manage (e.g., Protege [4], or SDA [5]).

Our approach to merging CPGs is sustained on the SDA language that represents the health-care procedural knowledge in the CPGs as clinical algorithms (CA) [6]. CAs can be found as part of CPGs [7], can be the result of a knowledge engineering process [8], or can be derived from health-care data [9]. The merging of CAs is based on the idea of finding out the basic knowledge units in each one of the CAs we want to merge, and then combine them into a single CA that gathers all the procedural knowledge scattered across the initial CAs. One of the main drawbacks of this approach is the need of complex and extensive expert knowledge to support one of the steps of the process. For each patient condition and action performed, this knowledge has to indicate what is the short-term expected evolution of that patient condition after applying that action. When there are several diseases involved, patient conditions and their evolution may be difficult to both evaluate and foresee by medical doctors. In such cases medical knowledge becomes complex because it cannot be found in CPGs, and large because the the number of alternative conditions of co-morbid patients may grow geometrically.

Here, we propose a machine learning procedure to carry out the merging of CAs that does not need the above mentioned complex and extensive expert knowledge. This procedure is sustained on the experience accumulated in health-care information systems about the treatment of comorbid patients.

A secondary objective of this work is show how the methodology of merging could be adapted to generate CAs from medical data. The result of the generation is suitable to be reused as knowledge for the merging of similar diseases.

In section 2 the merge of CAs is briefly introduced. Section 3 contains the description of the methodology to obtain CAs using medical data. Section 4 introduces two experiments with real data to generate different CAs. Finally, some conclusions are provided in section 5.

2   Merging CAs

Clinical algorithms are schematic models of the clinical decision pathway described in a CPG. These algorithms combine health-care actions with decision points in a sequential process that represents the long-term treatment of a particular disease [6]. The SDA model [5], SDA standing for State-Decision-Action, extends the above idea of CA with the concept of patient state to describe alternative conditions of the patient that require differential treatments, and also as a way of determining the feasible evolution of the patients across the states as they are treated of the disease. States are used as possible starting points in the application of the CA.

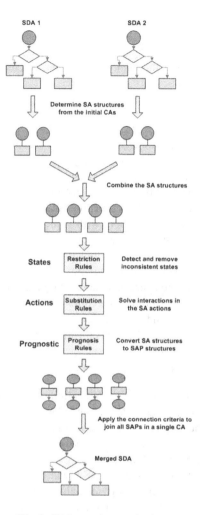

Fig. 1. SDA merging methodology

A methodology for merging CAs in the SDA model was proposed in [3]. Figure 1 summarizes this methodology as a sequence of transformations that go from the initial CAs to the merged CA.

This merging procedure is based on a limited number of domain variables. These variables are the basic components of the intermediate state-action (SA) and state-action-prognosis (SAP) structures that are basic to the merging procedure which consists of the following steps:

1. Determine SA structures from the initial CAs
2. Combine the SA structures
3. Detect and remove inconsistent states
4. Solve interactions in the SA actions

5. Convert SA structures to SAP structures
6. Apply the connection criteria to join all SAPs in a single CA

The next subsections presents the main concepts and discuss the issues of the merging procedure in more detail.

2.1 Domain Variables

The health-care terms in the CAs are expressed as state and action variables. State variables comprise all the important aspects that may be used to describe the condition of the patients, whereas the action variables are those elements that describe the activities carried out during the treatment of the patient (e.g. counsels, prescriptions, test requests, etc.). For example, the treatment of hypertension can be described with the state variables *DBP*[1], *SBP*[1], *cholesterol* (all of them with values *normal, high*, and *very high*), *bad lifestyle* and *obesity* (as Boolean variables); and with the action variables *drug prescription*, and *change lifestyle*. Detailed reference to the sort of drugs is also possible extending the action variables.

The merging process described here is based on the idea that the domain variables of the merged CA are taken from the domain variables of the respective CAs merged. From a health-care point of view, it could be the case that the resulting CA would require the use of new variables. This case has not been considered in this work.

2.2 SA and SAP Structures

The merging procedure consists in the decomposition of complex CAs in the basic SA structures. Each SA structure represents a knowledge of the sort *if-then* where the if part is the SA state, and the then part is the SA action. For example, the SA with S=*bad lifestyle, DBP high*, and A=*change lifestyle* proposes a change in the patient's lifestyle when a bad lifestyle and a high DBP is observed.

In the short-term evolution of a patient a SA can be extended to a SAP structure by introducing the expected evolution after an action A is performed on a patient with state S. In this paper, the concept short-term refers to the time between two consecutive patient-professional encounters. For example, the SAP with S=*bad lifestyle, DBP high*, and A=*change lifestyle*, and P=*DBP normal* indicates that the DBP of a patient with bad lifestyle will move from high to normal if the patient changes his lifestyle. On the contrary, long-term evolution refers to the medical changes happened between non consecutive encounters.

2.3 Combination of SA States

Two or more states can be combined to form a new state that contains all the state variables of the initial states. This sort of combination must not satisfy any of the constrains represented as a predefined set of restriction rules. A restriction rule is a subset of variables that cannot be observed simultaneously in the state of a patient. For

---

[1] DBP stands for Diastolic Blood Pressure
SBP stands for Systolic Blood Pressure.

example, if obesity and annorexy are state variables it is not be possible to have a patient state with both of them true.

Any state that satisfies a restriction rule is considered inconsistent, and the SA that contains it is removed from the set of SAs.

2.4   Combination of SA Actions

Some action variables as drug prescriptions may have interactions. These interactions and the way they are solved are expressed by means of substitution rules. A substitution rule is a tuple (S, A_1, A_2) where S represents a patient condition as a set of state variables, and A_1 and A_2 are respectively the actions before and after the interactions are solved in the medical context described by S. Both A_1 and A_2 are represented as sets of action variables. For example, the substitution rule:

({*diabetes, hypertension*}, {*beta-blocker*}, {*ACEi*[2]})

indicates that hypertensive diabetic patients that are treated with beta-blockers must change their medication to ACEi.

Every time two or more actions have to be combined in the merging procedure, all the substitution rules are used to detect and solve the feasible interactions.

2.5   Foreseeing Short-Term Prognosis

SA structures represent instant decisions where patients fulfilling S are treated as A indicates. However, CAs are essentially sequential and they reflect long-term decisions in time. In order to be able to convert SA structures in a CA we construct some intermediate structures called SAPs that capture the concept of short-term decision. A SAP (S, A, P) enlarges an SA (S, A) with the introduction of a predictive component that indicates what is the expected sort-term evolution of a patient in state S which is applied the treatment A.

A prognosis rule is defined as a tuple (S_p, A_p, S'_p) where S_p and S'_p are subsets of state variables, and A_p is a subset of action variables. This kind of rules represent the expert knowledge that is used to calculate the P components of SAP structures with the following procedure:

```
Procedure Transform SA to SAP (S, A, PR) is
    P := S;
    repeat
        for each (Sp, Ap, S'p) in PR do
            if (Sp ⊂ S) and (Ap ⊂ A)
            then P := (P \ Sp) ∪ S'p
        end;
    until P does not change;
    return (S, A, P);
end.
```

A prognosis rule (S_p, A_p, S'_p) is applicable to a SAP structure (S, A, P) if S and A include S_p and A_p, respectively. When the rule is applied, a new SAP structure

---

[2] ACEi stands for angiotensin-converting enzyme inhibitors.

(S, A, P') is obtained with $P' = (P \setminus S_p) \cup S'_p$. That is to say, all the variables in P that are in S_p are replaced by the variables in S'_p.

For each SA structure (S, A), the above procedure starts generating a SAP structure (S, A, S), and the prognosis rules (S_p, A_p, S'_p) are repeatedly applied, as previously indicated, until the SAP structure does not change. The order of application of prognosis structures is the one they are provided by the health-care expert.

2.6 Connecting SAPs

The last of the merging steps is the combination of all the obtained SAP structures in a single CA. The new CA is the result of applying the connecting steps described in [9] and it represents a treatment which is the combination of several treatments.

3 Using Medical Data to Obtain CAs

In the merging procedure explained in section 2, one of the main difficulties is to obtain medical knowledge from experts in the form of restriction, substitution and prognosis rules. An alternative to the introduction of this sort of knowledge by human experts is to use artificial intelligence algorithms to induce this knowledge from medical data. In this section we explore the induction of prognosis rules, and leave the induction of restriction and substitution rules for future work.

The induced rules will be used to transform SA structures into SAP structures as it is explained in section 2.5. In this new approach, unlike in the merging procedure introduced in section 2, the SA structures are not determined from the initial CAs, but automatically detected by the inductive algorithm in the data. The procedure to obtain the final CA is then composed of the following steps:

1. Determine the state and action variables from the medical data
2. Construct the data matrix
3. Find out the SA structures
4. Obtain the SAP structures
5. Apply the connection criteria to join all SAPs in a single CA

All these steps are described in the next subsections.

3.1 Medical Data

Medical data uses to come expressed in terms of specific vocabulary that is useful to represent concrete health-care activities, but which may lose their appropriateness when they have to be used in the description of CAs, where concepts have to be more general. This specific vocabulary is used to fill the data structures that contain the information about the treatment of patients. Data structures in this work are represented in figure 2 where each patient contains a list of encounters, each one with information about observations and health-care activities in the encounter.

Before the medical data is ready for the construction of CAs, the specific vocabulary in the data has to be converted to the state and action variables we want our final CA to

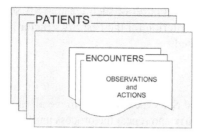

Fig. 2. Data model used in tests

contain. This process is made automatic with the introduction of transformation operators that allows the data to be filtered or generalized. These operators are of either the
form (*state variable, observation, condition, value*) or (*action variable, activity*). In the
first case all the observations that fulfill the condition in the data is converted to the state
variable with the given value. In the second case, all the activities in the data are converted to the action variable indicated. For example, (DBP, TAD, > 100, VERY-HIGH)
is an operator that converts all the TAD[3] values greater than 100 to the state Boolean
variable DBP-VERY-HIGH.

After the application of the transformation operators, all the state and action variables define a data matrix whose columns are the state variables twice and the action
variables once. The first state variables define the current state of the patient, the second
is the short-term evolution of the patient (i.e., prognosis variables) after the activities indicated by the action variables are applied. Each row of the matrix represents a different
encounter.

3.2 Obtaining SA Structures from the Data Matrix

An *action block* is each one of the different combination of values of the action variables
in the data matrix. Therefore, several encounters may share the same action block. We
express $(S_i = true)_A$ as the number of encounters whose state variable S_i is true in
the action block A, $(S_i = true)$ as the number of encounters whose state variable S_i
is true, and N_A as the number of encounters in the action block A. Then, equation 1 is
used to calculate the relevance of any state variable S_i in the definition of any action
block A.

This relevance is based on the idea that the most frequent a state variable appears in
an action block and the less it appears in other action blocks, the more relevant it is to
define the patients that are treated with that action block.

$$\alpha_i = \sqrt{\frac{(S_i = true)_A^2}{(S_i = true)N_A}} \tag{1}$$

Let S_κ^A be the set of the κ state variables with greater relevance for an action block
A. Then, (S_κ^A, A) describes a feasible SA structure representing N_A^κ encounters with
patients in state S_κ^A and with treatment A. N^κ is the number of encounters with patients

---

[3] TAD stands for the Spanish word for DBP.

in state S_κ^A and any treatment. For all of such feasible SA structures we define β_A^κ in equation 2 as the selection criterion for the best SA structure to be generated. The rest of structures are not generated.

$$\beta_A^\kappa = N_A^\kappa \left(\frac{N_A^\kappa}{N^\kappa}\right)^2 \kappa \tag{2}$$

If (S, A) is the generated SA structure, we remove all the encounters with patients in state S from the data matrix and repeat the process until the number of encounters in the matrix is reduced to a predefined percentage.

3.3   Obtaining Prognosis from Data

Each one of the SA structures obtained in the previous section has to be transformed into an SAP structure with the help of the data matrix. As described in section 3.1, the data matrix contains information about state, action, and prognosis variables. Given a SA structure (S, A), the expression $(P_i = true)_{(S,A)}$ is the number of encounters with patients in state S which receive treatment A and with prognosis variable P_i=true, and $(P_i = true)$ the number of encounters with prognosis variable P_i=true. These numbers are combined in equation 3 to calculate the prognosis capability of P_i in (S, A).

$$\gamma_i = (P_i = true)_{(S,A)} \frac{(P_i = true)_{(S,A)}}{(P_i = true)} \tag{3}$$

The prognosis P in the SAP structure contains the first κ prognosis variables with higher influence value γ, κ being a predefined parameter of the system.

3.4   Apply Connecting Conditions

Once the SAP structures are induced, they are joined in a CA with the connection criteria that is explained in [9].

4   Tests

To evaluate this work we propose an experiment with real data. Our objective is to obtain two different CAs from a dataset on the treatment of hypertension. The first CA is made to differentiate between when a patient requires drug treatment and when the patient requires changes in his lifestyle. On the contrary, the second CA is designed to show how to apply the different pharmacological treatments.

4.1   Data Description

The medical data used in this experiment has been provided by SAGESSA group[4]. It consists of a set of 28 patients with hypertension, equivalent to a total of 684 encounters with a minimum of 2 encounters and a maximum of 47 encounters per patient.

In the source data there are 100 different state variables, but only 43 appear more that 10 times, and 169 action variables, of which 134 correspond to drug prescriptions, 18 to request of new tests, and 17 to changes in the patient's lifestyle.

---

[4] http://www.grupsagessa.com/

4.2   Results

For the first CA we use the state variables *DBP, SBP, cholesterol, bad lifestyle* and *obesity*; and the action variables *drug prescription*, and *change lifestyle*. The medical data is filtered and generalized (see section 3.1) to obtain a data matrix with 131 out of 684 encounters, all of them containing at least one state variable, one action variable and one prognosis variable.

When the methodology described in section 3 is applied we obtain eight SA structures and eight prognosis rules. Figure 3 shows the resulting CA.

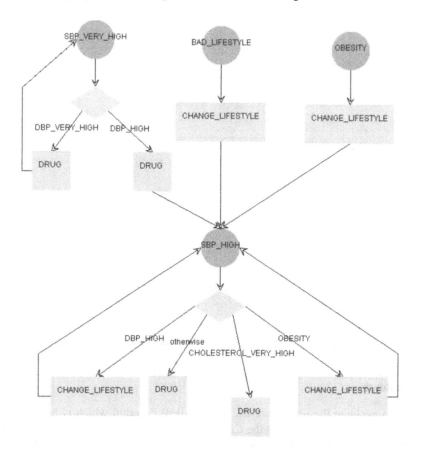

Fig. 3. CA of the general treatment of hypertension

The second CA takes all the state variables provided in the medical data, and the action variables *ACEi, Alpha-blockers, Beta-blockers, Diuretics, CCB*[5] and *ARB*[5]. After the filtering and generalization process the data matrix is reduced to 38 encounters. Figure 4 shows the resulting CA when we apply the second test.

---

[5] CCB stands for Calcium-Channel Blockers
ARB stands for Angiotensin-II Receptor Antagonists.

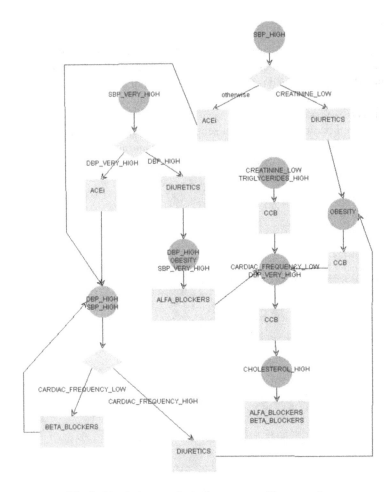

Fig. 4. CA of pharmacological treatment of hypertension

4.3   Evaluation

In order to evaluate the results we use an statistical criterion which is based on the classification of the encounters in the CAs. Table 1 shows the results of this evaluation where the number of encounters that can be included in at least one of the states of the CA appears as *classified*, and among them the number of encounters that also show a treatment equivalent to the sort of treatment indicated in the CA as *good classified*.

Attaining to these numbers and also to the CAs in figures 3 and 4 the main observations are:

- From a technical point of view both CAs are successful in the correct identification of the patient conditions with 83% and 92% of hits. The first CA also has a correct decision (74%) on the treatment the patients have to follow, where a pure random choice would have obtained just a 50%. In the second CA 50% of good classified

Table 1. Ratios of classification of encounters in the experiments

	First CA		Second CA	
Number of encounters	131	100%	38	100%
Classified	109	83%	35	92%
Well classified	97	74%	19	50%

encounters may be interpreted as a bad result, but pure random choice is only 14% and we must recall that the training sample is quite short to represent the treatment variability.

- From a medical perspective, two medical doctors have been asked to evaluate the CAs. Both agreed that the first CA describes correct medical actuations, except that the algorithm was very reluctant to prescribe drugs when the patient has SBP high. Both doctors recommend drugs in such cases. The expert analysis of the second CA is less positive, mainly because the state variables in the test were not all of the variables that should be used. This fact is also in close relation to the low good classified percentage discussed in the previous observation. In this CA, doctors also detected medical incoherences, which drove us to identify encounters reflecting odd medical actuations.

- From an intuitive analysis, the first CA shows interesting subtreatments as, for example, the stages of treatment when the SBP and the DBP are very high. In this case the treatment of the patient is centered in the reduction of DBP and then the SBP. Some other interesting facts are that, in case of bad lifestyle or obesity, the patient is recommended to change his lifestyle and then he is treated of a feasible high SBP.

5 Conclusion

This work shows that it is possible to obtain SA and SAP structures from data and to connect these structures in a CA. The quality of the results depends of the selection of the state and the action variables and also of the quality and the quantity of medical data available.

For the same dataset, it is possible to obtain alternative CAs each one representing a different point of view of the treatment, depending on the state and action variables selected.

Another issue is the quality of the medical data. Using data that is not representative of all the variability of the medical treatment we are trying to learn, may cause that the final CA is not medically correct. Sometimes, this fact may be used to detect the particularities of the treatments in a hospital or health-care irregularities.

In the studied cases, some problems as *Obesity* may require the application of actions that go beyond the following encounter or patient state. These long-term prognosis is not detected with the explained methodology.

The methodology has been tested at a structural level, but a further refinement is still required before the algorithm could be accepted by physicians or used in clinical practice.

Acknowledgments

This research has been partially funded by the EU project K4CARE (IST-2004-026968) and the Spanish National project HYGIA (TIN2006-15453).

References

1. Field, M.J., Lohr, K.N. (eds.): Clinical Practice Guidelines: Directions for a New Program, Institute of Medicine. National Academy Press, Washington (1990)
2. Shayegani, S.: Knowledge Modeling to Develop a Clinical Practice Guideline Ontology: Towards Computerization and Merging of Clinical Practice Guidelines, Master Thesis. Dalhousie University, Halifax, Nova Scotia, Canada (2007)
3. Real, F., Riaño, D.: Automatic combination of Formal Intervention Plans using SDA* representation model. In: Riaño, D. (ed.) K4CARE 2007. LNCS (LNAI), vol. 4924, pp. 75–86. Springer, Heidelberg (2008)
4. Shankar, R., Tu, S., Musen, M.: Use of Protege-2000 to Encode Clinical Guidelines, Stanford Medical Informatics Reports (2002),
   http://citeseer.ist.psu.edu/shankar02use.html
5. Riaño, D.: The SDA* Model: A Set Theory Approach. In: Proceedings of Computer Based Medical Systems 2007, Maribor, Slovenia, pp. 563–568 (2007)
6. Hadorn, D.C.: Use of algorithms in clinical guideline development. In: Clinical Practice Guideline Development: Methodology Perspectives. Agency for Health Care Policy and Research, Rockville, MD, Number 95-0009, pp. 93–104 (2005)
7. National Guideline Clearninghouse, http://www.guideline.gov/
8. Georg, G.: Computerization of Clinical Guidelines: an Application of Medical Document Processing. In: Proceedings of Computer Based Medical Systems 2007, Maribor, Slovenia, pp. 575–580 (2007)
9. Real, F., Riaño, D., Bohada, J.A.: Automatic Generation of Formal Intervention Plans Based on the SDA Representation Model. In: Proceedings of Computer Based Medical Systems 2007, Maribor, Slovenia, pp. 575–580 (2007)

OncoTheraper: Clinical Decision Support for Oncology Therapy Planning Based on Temporal Hierarchical Tasks Networks

Juan Fdez-Olivares[1], Juan A. Cózar[2], and Luis Castillo[3]

[1] Department of Computer Science and A.I, University of Granada, Spain
faro@decsai.ugr.es
[2] Pediatrics Oncology Service, Hospital Complex of Jaén, Spain
jcozarolmo@hotmail.com
[3] Department of Computer Science and A.I, University of Granada, Spain
L.Castillo@decsai.ugr.es

Abstract. This paper presents the underlying technology of *OncoTheraper*, a Clinical Decision Support System for oncology therapy planning. The paper is focused on the representation of oncology treatment protocols by a temporally extended, Hierarchical Task Networks (HTN) based knowledge representation as well as their interpretation by a temporal HTN planning process. The planning process allows to obtain temporally annotated therapy plans that support decisions of oncologists in the area of paediatrics oncology. A proof on concept aimed to validate this technology is also described. In addition, a service oriented architecture that supports the decision services of the system is proposed.

1 Motivation

The development of therapy planning systems [1,2,3] aimed to recommend predefined general courses of action to be applied to a patient, on the process of treating a disease, is an active research area in the general field of Clinical Decision Support Systems (CDSS) [4,5]. Decision support systems for therapy planning incorporate, on the one hand, a computerized representation of clinical protocols, also called computer interpretable clinical guidelines (CIGs)[6]: evidence-based operating procedures that physicians follow as a guide in order to perform clinical tasks as well as making clinical decisions. Most of the approaches in this field have focused on the development of languages and frameworks to support modeling, editing and representing CIGs [7], all of them based on "Task Networks Models" [6] (the knowledge represented follows a procedural scheme based on tasks and how they are decomposed into subtasks) where mechanisms to represent workflow patterns [8] that describe the process logic between subtasks are also included (mainly sequential, conditional, iterative and synchronization control structures). On the other hand, some systems [1,9,10] incorporate a reasoning process that is driven by the procedural knowledge encoded in protocols and, thus, interprets such representation by supporting clinical decisions made by experts. The great part of these approaches have centered on temporal constraints reasoning [9,10] aimed to validate constraints on a previously generated plan [3], but very little attention has been paid to the automated generation of therapy plans [2,11].

D. Riaño (Ed.): K4HelP 2008, LNAI 5626, pp. 25–41, 2009.

AI Planning and Scheduling (P&S)[12] seems to be the most adequate set of techniques to cover this aspect since it deals with the development of planning systems capable of interpreting a planning domain as a set of actions schemes (that might support the representation of a clinical protocol) and reasoning about them in order to compose a suitable plan (a sequence of actions) such that its execution reaches a given goal (to treat a patient) starting from an initial state (that might represent clinical data for a given patient profile). Concretely, HTN planning [13,14] becomes the most suitable AI P&S technique since it supports the modeling of planning domains in terms of a compositional hierarchy of tasks networks representing compound and primitive tasks by describing how every compound task may be decomposed into (compound/primitive) sub-tasks and the order that they must follow, by using different methods, following a reasoning process driven by the procedural knowledge encoded in its domain. These techniques have been successfully applied to real problems [15,16] but the main criticism received, regarding their application to therapy support in the medical domain [1], has been centered in their incapacity to represent and manage crucial temporal aspects needed in this domain, as well as lack of support for a flexible execution of plans so obtained. Indeed, this has been true until very recently [13], where HTN techniques have been enhanced with valuable temporal extensions that allow to cope with a very rich temporal representation, as well as to obtain plans that could be flexibly executed as they contain temporal constraints that can be adapted during plan execution.

Therefore, in this paper we will describe an application of temporal HTN planning techniques to both, represent computer interpretable oncology clinical protocols, and automatically generate personalized therapy plans for oncology patients, following a deliberative hierarchical planning process driven by the procedural knowledge presented in such protocols. The representation language that supports the description of such knowledge also allows to represent temporal constraints that are incorporated into the reasoning process in order to obtain temporally valid plans, suitable to be applied as oncology therapy plans. These techniques are the basis of the underlying technology of *OncoTheraper*, a CDSS that at present is being developed under the framework of a recently started R&D project (funded by the Regional Andalusian Government) participated by our research group together with the pediatrics oncology services (distributed in 6 different hospitals) of the Public Health System of Andalusia and two private companies (IActive Intelligent Solutions, a spin-off started up from our research group, and AT4Wireless).

The paper is structured as follows. First, the target domain application (pediatrics oncology) of this approach will be described. Then the main features of the representation language and the therapy planning process will be detailed. In addition, we will show a proof of concept performed in the Hospital Complex of Jaén (Spain) aimed to validate the representation, generation and visualization of oncology therapy plans (in close collaboration with oncologists). Last sections are devoted to describe the architecture of the system as well as some related work and conclusions.

2   Domain of Application

The work here presented is focused on the paediatrics oncology area, in which health assistance (and particularly therapy planning) is based on the application of oncology treatment protocols: a set of operating procedures and policies to be followed in both stages, treatment and monitoring of a patient. The main goal of an oncologist when planning a treatment is to schedule chemotherapy, radiotherapy and patient evaluation sessions. These sessions should be planned following different workflow patterns [8], included in the protocol, that specify tasks at different levels of abstraction, including sequential, conditional and iterative control flow logic constructs. Furthermore, sessions are organized as cycles of several days of duration where every cycle includes the administration of several oncology drugs. Additionally, drugs are administrated following different *administration rules* regarding their dosage and duration. Monitoring sessions must also be considered and scheduled according to protocol guidelines.

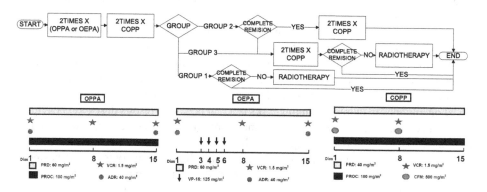

Fig. 1. A general schema of Hodgkin's Disease Clinical Protocol. The representation followed to show the periodical temporal patterns for chemotherapy cycles (OPPA, OEPA and COPP) is literally copied from the protocol specification.

Figure 1 shows an example of a concrete oncology treatment protocol (the one followed at present for planning the treatment of Hodgkin's disease [17] and elaborated by the Spanish Society on Pediatrics Oncology). A general schema of the treatment workflow process indicated in such clinical protocol is outlined in the flow-chart diagram of this figure. First a child must receive two chemotherapy cycles (of type OPPA or OEPA, depending on the genre) and another two cycles of type COPP. If a complete remission of the tumour is not achieved by patients of *Group1* then radiotherapy sessions must start. If the stratification group[1] (decided by the oncologist) is either *Group2 or Group3* two more COPP cycles must be administrated. In case of a patient of *Group3*, additional radiotherapy sessions must be administrated when a complete remission of the tumour is not detected. Temporal patterns to administrate every type of chemotherapy cycle are shown below the flow-chart of Figure 1. For example, a cycle of type OPPA takes 15 days, the rules to administrate a cycle of type OPPA state that the drugs PRD and PROC must be administrated every day (dosage is also specified), VCR has to be administrated the first, eighth and last day, and ADR the first and last day (OEPA and

---

[1] Patients that receive a given protocol are initially stratified in a group depending on several criteria like the size of their tumour.

COPP patterns should be interpreted in a similar way). In addition, start times for every chemotherapy cycle must be separated at least 28 days, and an evaluation session has to be scheduled previously to the start of every cycle.

Therefore, in most therapies, actions concerning drugs administration and patient evaluation have to be performed according to a set of temporal constraints describing their relative order, and the delays between them. Additionally, in many cases, actions must be repeated at regular (i.e. periodic or following a repetition pattern) times. Furthermore, it is also necessary to carefully take into account the (implicit) temporal constraints derived from both, the hierarchical decomposition of actions into their components, and from the control-flow of actions in the clinical protocol [10]. All these rules, tasks and decisions vary depending on a given patient profile and may change as the treatment is going on.

At present, planning a therapy in the hospital services that concern to this work (pediatrics oncology services in the Public Health System of Andalusia) is done by hand, that is, thought it is possible to access patient's medical information in Electronic Health Records (EHR), there is no tool to support decisions made while planning the treatment and monitoring sessions of patients. The deployment of a decision support system to assist oncologists in therapy planning tasks is a real need that results in several benefits: workload of oncologists will be reduced and more time might be dedicated to personal assistance to patients (improving quality of health delivery), patient safety is augmented by automating administration rules, and efficiency of health delivery is increased since resource coordination and usage will be supported by an automated planning process that incorporates representation and reasoning about time and resources.

The following section is devoted to describe how tasks concerning the stages of treatment and monitoring performed by oncologists, their internal process logic, and the temporal constraints to be observed during a treatment, can be represented by a temporally extended, HTN-based knowledge representation scheme. First, the main features of the HTN P&S system [13,15], capable of managing such representation and used as the core technology to support oncologists' decisions on therapy planning will be summarized, then knowledge representation as well as planning and temporal reasoning aspects will be detailed.

3   Representing and Reasoning about Oncology Protocols

The AI Planning and Scheduling system used in this work to operationalize oncology protocols has been developed by our research group and, furthermore, has already been applied to other practical problems [15]. It uses as its planning domain and problem description language an HTN extension of PDDL (Planning Domain Description Language)[18], a language used by most of well known planners that allows to represent non-hierarchical planning domains as a set of actions with typed parameters, preconditions and effects. Actions' preconditions represent the conditions that must hold in order to execute a given action, while actions' effects are intended to represent changes in the world by defining which facts are asserted and retracted by the execution of an action. Numerical functions are also allowed in both, preconditions and effects (what provides support to compute, for example, the duration of a

```
(:durative-action AdminDrug
 :parameters (?p - Patient ?ph - Drug ?ds ?dur - number)
 :duration (= ?duration ?dur))
 :condition (patient_ok ?p)
 :effect (increase (total_dosage ?p ?ph) ?ds))


(:task Protocol
 :parameters (?p - Patient ?date - Date)

 (:method Group1
  :precondition (= (group ?p) Group1)
  :tasks (
   (eval_patient ?p)
   [((and (= ?duration 360)(>= ?start ?date))(ChemoTherapy ?p))
       (RadioTherapy ?p)]
   (eval_patient ?p)))

 (:method Group2
  :precondition (= (group ?p) Group2)
  :tasks (
   (eval_patient ?p)
   [((and (= ?duration 360)(>= ?start ?date))(ChemoTherapy ?p))
       (RadioTherapy ?p)]
   (eval_patient ?p)
   [((= ?duration 360)(ChemoTherapy ?p))
       (RadioTherapy ?p)]
   (eval_patient ?p))))
```

Fig. 2. HTN-PDDL concepts: A *primitive task* and a *compound task* with two *decomposition methods*

drug-administration action depending on patient conditions) and, therefore, it is also possible to represent either consumable or discrete numerical resources (for example, the total drug dosage received by a patient, see *:durative-action* in Figure 2). Concretely, primitive tasks of our HTN−PDDL extension, are encoded as PDDL 2.2 level 3 durative actions (allowing to represent temporal information like duration and start/end temporal constraints, see [13] for details). In addition, HTN methods used to decompose compound tasks into sub-tasks include a precondition that must be satisfied by the current world state in order for the decomposition method to be applicable by the planner. For example, (:task Protocol in Figure 2 describes two alternative courses of action depending on the group a patient belongs to, by using two different decomposition methods.

The basic planning process (shown in Figure 3) receives as input a set of facts that represent an initial state of the world (that describes the context of the health-care treatment, including patient's current state) as well as a goal, described as a partially ordered set of tasks that need to be carried out. Then it follows a state-based forward HTN planning algorithm that decomposes that top-level set of tasks and its sub-tasks by selecting their decomposition methods according to the current state and following the order constraints posed in tasks decomposition schemes as a search-control strategy. Therefore, it explores the space of possible decompositions replacing a given task by its component activities (that may be either primitive or compound), until the initial set of tasks is transformed into a set of only primitive actions that make up de plan.

- Set $\mathcal{A}$, the agenda of remaining tasks to be done, to the set of high level tasks specified in the goal.
- Set $\Pi = \emptyset$, the plan.
- Set $\mathcal{S}$, the current state of the problem, to be the set of literals in the initial state.
  1. Repeat while $\mathcal{A} \neq \emptyset$
     (a) **Extract** a task t from $\mathcal{A}$
     (b) if t is a primitive action, then
         i. If $\mathcal{S}$ satisfies t preconditions then
            A. Apply t to the state, $\mathcal{S} = \mathcal{S} + additions(t) - deletions(t)$
            B. Insert t in the plan, $\Pi = \Pi + \{t\}$
            C. **Propagate-Temporal-Constraints(Π)**
         ii. Else **FAIL**
     (c) if t is a compound action, then
         i. If there is no more decomposition methods for t then **FAIL**
         ii. **Choose** one of its decomposition methods of t whose preconditions are true in $\mathcal{S}$ and map t into its set of subtasks $\{t_1, t_2, \ldots\}$
         iii. Insert $\{t_1, t_2, \ldots\}$ in $\mathcal{A}$.
  2. **SUCCESS**: the plan is stored in Π.

Fig. 3. A rough outline of an HTN planning algorithm showing the point at which temporal constraints are propagated (step 1(b)iC)

This forward search process makes the planner to know, at every step in the planning process, the current context of the health-care treatment, including patient's current state. Concretely, this context-awareness is specially important when preconditions of both primitive actions and methods are evaluated on the current state (respectively steps (b).i and (c).ii in Figure 3) , what allows to incorporate significant inferencing and reasoning power as well as the ability to infer new knowledge by requesting information to external hospital information services. In this sense, the planner uses two mechanisms addressed to represent as well as support oncologists decision-rules concerning issues like conserving patient safety on the administration of drugs.

On the one hand, *deductive inference tasks* of the form (:inline <p> <c>) may be fired in the context of a decomposition scheme, when the logical expression <p> is satisfied by the current treatment state, providing additional bindings for variables or asserting/retracting literals into the planner's knowledge base, depending on the logical expression described in <c>. These tasks can be used (as shown for example in Figure 4) to dynamically compute, depending on the current health-care context, the dosage an duration of drugs administration (from functions that define either the intensity of dosage or the time-rate, depending on the body surface of a patient). On the other hand, *abductive inference rules* of the form (:derived <lit> <expr>) allow to infer a fact <lit> by evaluating <expr>, that may be either a more complex logical expression or a Python script that both, binds its inputs with variables of <lit>, and returns information that might be bound to some of the variables of <lit>. For example, a derived literal might be used to infer whether a patient is in an correct state, from a complex expression including all the necessary conditions that enable the administration of a given drug (see derived literal on Figure 4). This literal might then be used as a precondition of an action that represents the task of administrating a drug (as shown in Figure 2).

3.1 Representing Workflow Patterns

Compound tasks, decomposition methods and primitive actions represented in a planning domain mainly encode the procedures, decisions and actions that oncologists must

```
(:derived (patient_ok ?p)
 (and (> (leucocites ?p) 2000)
      (> (neutrophils ?p) 500) ))

(:task ChemoTherapy
 :parameters (?p - Patient)

 (:method repeat
  :precondition (> (NRep ?p VCR) 0)
  :tasks (
   (:inline () (decrease (NRep ?p VCR)))
   (:inline () (assign ?dosage (* (surface ?p) (intensity ?p))))
   (:inline () (assign ?dur (* (surface ?p) (time_rate ?p))))
   (AdminDrug ?p VCR ?dosage ?dur)
   (ChemoTherapy ?p)))

 (:method base_case
  :precondition (= (NRep ?p VCR) 0)
  :tasks ()))
```

Fig. 4. HTN-PDDL concepts: A *derived literal*, and a task with a recursive decomposition scheme, including *inline tasks*

follow, according to a given oncology protocol, when they deal with a treatment on a given patient. More concretely, the knowledge representation language as well as the planner are also capable of representing and managing different workflow patterns present in any of such protocols (also present, on the other hand, in most CIGs formalisms [7,6]). A knowledge engineer might then represent control structures that define both, the execution order (sequence, parallel, split or join), and the control flow logic of processes (conditional and iterative ones). For this purpose the planning language allows sub-tasks in a method to be either sequenced, and then they appear between parentheses (T1,T2) , or splitted, appearing between braces [T1,T2]. Furthermore, an appropriate combination of these syntactic forms may result in split, join or split-join control structs. For example, decomposition methods of the main task `Protocol` (Figure 2) describe that chemotherapy and radiotherapy sessions must be executed in parallel, but they must be synchronized with both a previous (split) and a later (join) evaluation of the general state of a patient (issues about temporal information included in the decomposition scheme shown will be detailed later).

Conditional and iterative control constructs can also be represented as task decomposition schemes that exploit the main search control techniques implemented by the planner. Briefly, a general process p that contains a conditional struct *if c then p1 else p2* can be represented as a task decomposition scheme as the one shown in the task `Protocol` (Figure 2), that encodes a conditional structure based on the stratification group of a patient. This decomposition scheme describes that if a condition c (a patient belongs to *Group1*) holds in the current health-care context, then apply (`:method Group1`) else apply (`:method Group2`).

On the other hand, a general process p that contains an iterative struct *while c p1* may be represented as a task decomposition scheme as the one shown in the task `Chemotherapy` (Figure 4). This decomposition scheme describes that the primitive task `AdminDrug` should be repeatedly performed while the number of repetitions prescribed for the drug VCR (Vincristine) is greater than 0.

3.2   Representing and Reasoning about Temporal Constraints

Furthermore, our HTN domain description language as well as the planning algorithm support to explicitly represent and manage time and concurrency at every level of the task hierarchy in both compound and primitive tasks, by allowing to express temporal constraints on the start or the end of an activity. Any sub-activity (either task or action) has two special variables associated to it, `?start` and `?end`, that represent its start and end time points, and some constraints (basically `<=`, `=`, `>=`) may be posted on them (it is also possible to post constraints on the duration with the special variable `?duration`). In order to do that, any activity may be preceded by a logical expression that defines a temporal constraint as it is shown in (`:task Protocol` (Figure 2), where the duration of any chemotherapy session (an sub-tasks included in its decomposition) is constrained to 360 hours (15 days). The beginning of chemotherapy (in any of the two alternative courses of action) is constrained to start not earlier than a given date.

This temporal knowledge can be managed by the planning process thanks to the handling of metric time over a Simple Temporal Network (STN), a structure (X, D, C) such that X is the set of temporal points, D is the domain of every variable and C is the set of all the temporal constraints posted (See [13] for more details). In our case, a plan is deployed over a STN following a simple schema: every primitive action a_i included in a plan owns two time points $start(a_i)$ and $end(a_i)$, and every compound task t_i decomposed during the planning process generates two time points $start(t_i)$ and $end(t_i)$ which bound the time points of its sub-tasks. These temporal constraints are encoded as absolute constraints with respect to the absolute start point of a STN. All the time points share the same domain $[0, \infty)$, but it is important to note that the constraints in C (described in the planning domain) provide support to describe flexible temporal constraints, by defining earliest and latest execution times for start/end points associated to every task or action. For example, it is possible to encode constraints of the form ((`and (>= ?start date1) (<= ?start date2)) (t)`) what provides flexibility for the start time of `t`'s execution, indicating that `t` should start neither earlier than `date1` nor later than `date2`.

Every time that a compound or primitive task is added to the plan, all the time points and constraints of the STN are posted, propagated and validated automatically, observing both the implicit (derived from qualitative order constraints) and explicit (derived from quantitative constraints described in the domain) temporal constraints defined in any decomposition scheme. This temporal representation, on the one hand, provides enough expressivity power to truly represent workflow schemes such as sequence, parallel, split and join, since during the planning process our planner is capable of inferring quantitative temporal constrains from the qualitative ordering constraints expressed in decomposition methods. On the other hand, time points of subtasks of any task `t` with temporal constraints are embraced by the time points of `t`, what means that subtasks inherit the constraints of their higher-level task. This allows to represent and reason about temporal constraints derived from hierarchical decompositions, a strong requirement of any system devoted to support therapy planning (as stated in [10]).

3.3   Representing Periodic Tasks and Temporal Constraints

The HTN planner is also able to record the start and end of any activity and to recover these records in order to define complex synchronization schemes between either tasks or actions, as relative constraints with respect to other activities. This mechanism is used to encode synchronization of tasks that correspond to repetitive periodic patterns. The first step is the definition, by assertion, of *temporal landmarks* that signal the start and the end of either a task or an action (Figure 5). These landmarks are treated as PDDL functions (predicates that represent functions which when evaluated return a value or an object, in this case, a time-point of the STN) that are associated to the time points of the temporal constraints network.

```
(:durative-action AdminDrug
 :parameters (?p - Patient ?ph - Drug ?ds ?dur - number)
 :duration (= ?duration ?dur)
 :condition (patient_ok ?p)
 :effect (and (increase (total_dosis ?p ?ph) ?ds))
              (assign (last-admin ?p ?ph) ?end))

(:task A3
 :parameters (?p - Patient ?ph - Drug)
 (:method A3
  :precondition (...)
  :tasks (((= ?start (last-admin ?p ?ph)) (b)))))
```

Fig. 5. Generating and recovering a temporal landmark

These landmarks are asserted in the planner's current state, and later on, they may be recovered and posted as constraints of other tasks in order to synchronize two or more activities. For example, Figure 5 shows how to recover a temporal landmark that restricts action b to start exactly at the same time than action AdminDrug ends. That is, the primitive action AdminDrug assigns, by means of its effects, the end time-point (represented by the variable ?end) to the function (last-admin ?p ?ph), and asserts this temporal landmark in the current state. The task A3 includes in its decomposition method a temporal constraint that synchronizes the sub-task b with the end point of AdminDrug, using the temporal landmark previously recorded.

In particular, thanks to the expressive power of temporal constraints networks and to the mechanism explained so far, a planning domain designer may explicitly encode in a problem's domain all of the different orderings included in Allen's algebra [19] between two or more tasks, between two or more actions or between tasks and actions. Furthermore, temporal landmarks are an excellent resource in order to express different kinds of periodic patterns to be followed by temporal constraints, a strong requirement of clinical protocols, particularly oncology clinical protocols. For example, Figure 6 shows a refined description of the Chemotherapy task previously shown in Figure 4 that combines temporal landmarks management and recursive decompositions in order to specify that the administration of VCR must be always preceded by a delay of 24 hours, and must be repeated a number of times defined by the function (NRep ?p VCR). Additionally, recall that in Figure 2 the task Chemotherapy was constrained to a duration of 360 hours (15 days), and, since the planning process allows subtasks to

```
(:task ChemoTherapy
 :parameters (?p - Patient)

 (:method repeat
  :precondition (> (NRep ?p VCR) 0)
  :tasks (
   (:inline () (decrease (NRep ?p VCR)))
   (:inline () (assign ?dosage (* (surface ?p) (intensity ?p))))
   (:inline () (assign ?dur (* (surface ?p) (time_rate ?p))))
   ((and (>= ?start (last-admin ?p VCR)) (= ?duration 24)) (Delay ?p VCR))
   ((and (= ?duration ?dur)) (AdminDrug ?p VCR ?dosage ?dur)))
   (ChemoTherapy ?p))

 (:method base_case
  :precondition (= (NRep ?p VCR) 0)
  :tasks ())))
```

Fig. 6. A chemotherapy cycle that uses temporal landmarks management and recursive decompositions

inherit constraints of higher-level tasks, all the actions of this chemotherapy cycle must be executed in an interval of 15 days (360 hours).

4 Technology Validation

Considering the previous description, a proof of concept of this technology has been carried out in collaboration with expert oncologists in the Hospital Complex of Jaén (Spain). During this proof, the oncology treatment protocol for the Hodgkin's disease has been encoded in the planning language above described, in a knowledge elicitation process based on interviews with experts. The domain includes six compound tasks, 13 methods, 6 primitive tasks and the file contains more than 400 lines of code[2]. During the proof of concept the planner received the following inputs: a planning domain, representing this protocol; an initial state representing some basic information to describe a patient profile (stratification group, age, sex, body surface, etc.) as well as other information needed to apply administration rules about drugs (dosage, frequency, etc.); and a high-level task representing the goal (apply the protocol to the patient) with temporal constraints representing the start date of the treatment plan. Experiments performed were aimed to validate both the knowledge represented (Hodgkin's protocol) and therapy plans generated. Oncologists played a validation role (based on interviews) in both stages, knowledge elicitation and representation, and plan validation. Their feedback informed us that therapy plans were accurate with respect to the protocol's guidelines encoded in the planning language. However, they also commented that a complete validation should include how these plans are executed, what is our goal in our current research work (as explained in conclusions). Plans generated represent therapy plans tailored to a given patient profile, and they allow to represent therapies of several months of duration, including tens of temporally annotated actions with start/end constraints. Figure 7 shows a therapy plan with 51 actions and 6 months of duration obtained for

---

[2] Available on http://decsai.ugr.es/~faro/Hodgkin/index.html

ID	Task Name	Duration	Start	Finish
1	**Hodgkin Group3 Female**	**24 hrs**	**Fri 07/12/07**	**Sat 08/12/07**
2	PreviousEval	1440 mins	Fri 07/12/07	Sat 08/12/07
3	**CYCLE OPPA**	**360 hrs**	**Sat 08/12/07**	**Sun 23/12/07**
4	AdminDrug VCR 0.866025 IV	1440 mins	Sat 08/12/07	Sun 09/12/07
5	AdminDrug VCR 0.866025 IV	1440 mins	Sat 15/12/07	Sun 16/12/07
6	AdminDrug VCR 0.866025 IV	1440 mins	Sat 22/12/07	Sun 23/12/07
7	AdminDrug PRD 34.641014 Oral_3_daily	21600 mins	Sat 08/12/07	Sun 23/12/07
8	AdminDrug PRC 57.735027 Oral_3_daily	21600 mins	Sat 08/12/07	Sun 23/12/07
9	AdminDrug ADR 23.094009 IV_Perfusion	1440 mins	Sat 08/12/07	Sun 09/12/07
10	AdminDrug ADR 23.094009 IV_Perfusion	1440 mins	Sat 22/12/07	Sun 23/12/07
11	PreviousEval	1440 mins	Mon 07/01/08	Tue 08/01/08
12	**CYCLE OPPA**	**360 hrs**	**Tue 08/01/08**	**Wed 23/01/08**
20	ResponseEval	1440 mins	Wed 23/01/08	Thu 24/01/08
21	**CYCLE COPP**	**360 hrs**	**Tue 05/02/08**	**Wed 20/02/08**
22	AdminDrug VCR 0.866025 IV	1440 mins	Tue 05/02/08	Wed 06/02/08
23	AdminDrug VCR 0.866025 IV	1440 mins	Tue 12/02/08	Wed 13/02/08
24	AdminDrug PRD 23.094009 Oral_3_daily	21600 mins	Tue 05/02/08	Wed 20/02/08
25	AdminDrug PRC 57.735027 Oral_3_daily	21600 mins	Tue 05/02/08	Wed 20/02/08
26	AdminDrug CFM 288.675140 IV_Perfusion_30min	1440 mins	Tue 05/02/08	Wed 06/02/08
27	AdminDrug CFM 288.675140 IV_Perfusion_30min	1440 mins	Tue 12/02/08	Wed 13/02/08
28	PreviousEval	1440 mins	Wed 20/02/08	Thu 21/02/08
29	**CYCLE COPP**	**360 hrs**	**Tue 04/03/08**	**Wed 19/03/08**
36	ResponseEval	1440 mins	Wed 19/03/08	Thu 20/03/08
37	**CYCLE COPP**	**360 hrs**	**Tue 01/04/08**	**Wed 16/04/08**
38	AdminDrug VCR 0.866025 IV	1440 mins	Tue 01/04/08	Wed 02/04/08
39	AdminDrug VCR 0.866025 IV	1440 mins	Tue 08/04/08	Wed 09/04/08
40	AdminDrug PRD 23.094009 Oral_3_daily	21600 mins	Tue 01/04/08	Wed 16/04/08
41	AdminDrug PRC 57.735027 Oral_3_daily	21600 mins	Tue 01/04/08	Wed 16/04/08
42	AdminDrug CFM 288.675140 IV_Perfusion_30min	1440 mins	Tue 01/04/08	Wed 02/04/08
43	AdminDrug CFM 288.675140 IV_Perfusion_30min	1440 mins	Tue 08/04/08	Wed 09/04/08
44	PreviousEval	1440 mins	Wed 16/04/08	Thu 17/04/08
45	**CYCLE COPP**	**360 hrs**	**Tue 29/04/08**	**Wed 14/05/08**
52	EvalRemission	1440 mins	Wed 14/05/08	Thu 15/05/08

Fig. 7. A therapy plan that follows the Hodgkin's disease protocol for a female patient of Group3

a female patient of stratification group Group3. For each administration action in the plan the following information is shown: drug name, dosage and administration mode; duration, and start and end dates. Some chemotherapy cycles are shown as a summary and their sub-actions are not shown for simplicity purposes. The protocol also states different types of evaluation sessions that are included in the plan.

Therapy plans are also represented in a standard XML representation and may be displayed as Gantt charts in standard tools devoted to project management (like MS

Project, see Figure 8). The plan shown in Figure 8 has been obtained after an automated postprocessing of the output of the planner, in order to friendly show the tasks of the plan (left hand side of the figure) as well as their temporal dimension as a Gantt chart (right hand side). The visualization of the tasks in a MS Project display allows to show tasks in a *Work Breakdown Structure* including different outline levels (either *summary tasks* as OPPA CYCLE or *standard taks* as AdminDrug), that may be collapsed or deployed as shown in the figure. It is worth to note that this structure is managed by the planner from the knowledge encoded in the domain, taking advantage of the possibility of encode additional special features in a procedural knowledge representation as the one supported by our planning language. On the other hand the Gantt chart visualization offers an outline of how these tasks are correctly arranged following the periodic rules of every type of chemotherapy administration. Contrasted with oncologists, only the generation and visualization in few seconds of a therapy plan (in this case only chemotherapy sessions are displayed) is considered of great help, since it saves a lot of time in their current decision making process, because oncologist have to take into account too many detailed constraints and tasks that, on the other hand, we have shown that can be accurately represented in a temporally extended HTN representation.

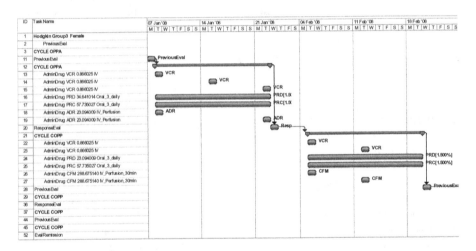

Fig. 8. A temporally annotated and automatically generated therapy plan represented as a gantt chart

This proof of concept has served to validate the technology here presented as the basis to support both, clinical processes and decisions in the field of (pediatrics) oncology therapy planning. At present, a Clinical Decision Support System for oncology therapy planning is being developed in the framework of *OncoTheraper*, a recently started R&D project (funded by the Regional Andalusian Government) participated by our research group together with the pediatrics oncology services (distributed in 6 different hospitals) of the Public Health System of Andalusia and two private companies (IActive Intelligent Solutions, a spin-off started up from our research group, and AT4Wireless). The architecture of this system is outlined in the following section.

5 Proposed Architecture

OncoTheraper is designed to be supported by a Service Oriented Architecture (SOA) aimed to support ubiquitous access (through appropriated visual interfaces) to the decision services offered at any stage of the care providing life-cycle (protocol modeling, diagnosis, treatment and monitoring) by its main components (see Figure 9): an oncology protocols server, a therapy planning server and an intelligent monitoring server.

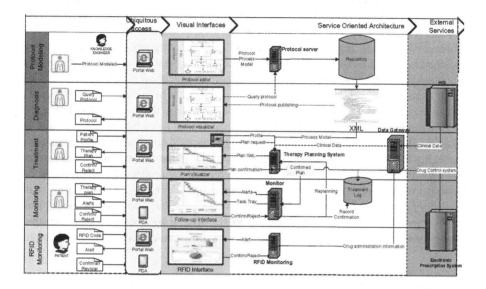

Fig. 9. Proposed architecture of OncoTheraper

The *Protocols Server* is designed to store the computerized oncology protocols in the representation language described in this work, this will allow to represent protocols in a format ready to use by the planner. For this purpose, it is planned to use a state-of-art CIG modeling tool (like the one presented in [20] and to develop a back-end to translate the formally modeled protocol into the planning language. The architecture's design is aimed to guarantee that knowledge engineers will only take part in the protocol modeling stage since, once developed the appropriate back-end and validated the oncology protocols modeled, the system will require only the interaction with oncologists.

The *Therapy Planning Server* is based on the techniques described in this work. In this sense one of the main challenges in the development of this service (also aligned with other systems [21]) is related to the integration with legacy clinical information systems already existing in the working environment of oncologists, particularly those related with both how to translate the information stored in Electronic Health Records (EHR) into the planning representation language, and how to store the plans obtained by the planner in EHRs following a standard representation for clinical information. Most hospitals incorporate in their clinical information systems a *drug stock control system* that is already being used by oncologists to send drug administration orders. Indeed, the opinion of oncologists is that the output of the therapy planning system here presented might be of great help if it were integrated with the input to such system.

The *Intelligent Monitoring Server* is designed to interpret and execute the therapy plans generated by the therapy planning server. These plans contain actions that represent activities as well as decisions an oncologist should follow, and they are deployed over a STN used to represent time intervals that constraint both start and end execution times of actions. Therefore, at the beginning of the execution of a therapy plan, actions, temporal constraints and facts that represent preconditions and effects of actions are consistent with respect to the initial conditions expressed in the therapy planning problem. Additionally, regarding plan execution, a monitoring process has already been developed (applied to a different domain application [15] that, nevertheless, shares this same plans representation) that guarantees the correct execution of actions, thus avoiding for example the activation of actions once they have been finished. However, as the plan execution is progressing, inconsistencies may arise that could affect either the temporal dimension of the plan or actions' preconditions. In such cases a rescheduling process might be carried out devoted to rearrange temporal constraints, by checking the consistency of the underlying plan's temporal network. In the case that a consistent temporal network couldn't be found, an automated replanning process (based on the same planning process here described) might be triggered in order to readapt the therapy plan to new circumstances. In addition, in order to improve patient safety, one of the companies involved in the project will integrate a *RFID Monitoring system* to guarantee that the drug orders defined by the therapy planning system are correctly administrated to patients.

6   Related Work

The planning process and representation so far described present some advantages with respect to current state of the art techniques devoted to therapy planning that are worth to note. Firstly, the representation and reasoning about temporal constraints of our approach allows to simultaneously validate temporal constraints while generating therapy plans (plan generation and temporal constraint management are interleaved). Most approaches [1] are only focused in one side of the problem of therapy planning, since they only pay attention on how to manage temporal constraints of actions, and neglect aspects related to how automatically generate sequences of actions with temporal constraints. Very few [9,3] face the problem of plan generation, but it is carried out following a non-deliberative process, based on the static adaptation of skeletal-plans, very close to case-based planning, that is not interleaved with temporal constraints reasoning. Instead of this, these approaches are based on a batch process that firstly generates a complete plan and then analyzes its temporal constraints, what affects negatively to the efficiency of the overall process, as well as to important reasoning aspects like the loss of backtracking points (which are lost when a plan is completely generated) or the impossibility of using the causal rationale of the plan as a guide to propagate constraints (as is the case of our planner [13]). These reasoning aspects are specially important when plans have to be readapted due to new circumstances arisen during the treatment stage.

Therefore, this approach should not be considered only as a new way to represent therapies, in addition it provides support to leverage the plan management life-cycle since it only requires knowledge engineer intervention at the beginning (in the protocol modeling stage). On the contrary, the knowledge and plan management life-cycle of other approaches devoted to therapy plan management (like Asbru [9] or Glare [10]) require specialized human intervention (either knowledge engineers or trained medical staff) when tailoring a therapy plan from an initial protocol scheme to a given patient profile (what is directly done by our therapy planning system). These approaches are mainly focused on the verification of therapy plans with temporal constraints (apart from providing very expressive CIGs representation formalisms) and we have shown that our temporal representation and reasoning is as expressive as the one used in Asbru or Glare. Furthermore, the process performed by these approaches to temporal constraints verification could be used at execution time in order to revise possible temporal inconsistencies (like a delay in the administration of a drug), but there are circumstances in which the actions included in a therapy plan (and not only temporal constraints) must be partial or completely readapted (for example, when a patient's stratification group changes since his/her tumour size does not progress as expected). In such cases our approach might use the same planning process to automatically readapt the therapy plan, by shifting more detailed decisions to the planner and reducing the workload of oncologists, as opposite to current approaches that always need to readapt from the scratch.

7 Conclusions

In this work we have presented the architecture of OncoTheraper, a CDSS whose underlying technology is based on an AI P&S system that uses temporal Hierarchical Task Networks (HTN) planning techniques in order to automatically and dynamically generate personalized therapy plans for pediatrics oncology patients, following a deliberative hierarchical temporal planning process, driven by the procedural knowledge described in oncology protocols. The technology has been validated in a proof of concept at the Pediatrics Oncology Service of the Hospital Complex of Jaén (Spain), where oncologists were involved in the validation of both the knowledge represented and the oncology therapy plans obtained.

From the health assistance point of view, this approach presents some benefits: on the one hand, oncologists recognize that their workload might be reduced in benefit of the patient (they spend hours in planning an accurate therapy while our system obtains the same therapy plans in few seconds), thus improving health delivery quality. On the other hand, patient safety might be augmented since the recommended actions to administrate drugs are based on an automated planning process and, in addition, the final drug administration orders may be safely controlled by a RFID monitoring system at the health point care.

From a technical point of view, the knowledge representation language here described has been proved to be expressive enough to represent clinical processes as well as complex, periodic temporal constraints that appear in oncology protocols. In addition, the usefulness of therapy plans obtained have been recognized by oncologists, since these plans fit accurately to the guidelines of protocols. However, it is necessary to recognize that there is not yet a serious evaluation about the impact on patients from a clinical point of view, as the full system has not been deployed. Indeed, this is the main goal of the current research project where many remaining issues are subject to further development. For example, the development of appropriated ontology-based techniques [21] for the integration of the therapy planning system with legacy clinical information systems already existing in the working environment of oncologists.

Finally, we cannot neglect the use of knowledge engineering techniques in order to support the process of representing oncology protocols in our planning language. It is well known the proliferation of standard languages and frameworks for modeling and editing CIGs [6,20]. As explained in the introduction and shown thorough this paper our planning language embodies most of the features of such languages. Indeed, our next planned step is to represent oncology clinical protocols into one of these standard schemes and to develop a fully automated translation process from such representation to our planning language, thus allowing to automatically generate, execute and monitor treatment plans from a standard representation, keeping to the minimum the intervention of knowledge engineers in the whole knowledge management life-cycle.

Acknowledgements

This work has been partially supported by the Andalusian Regional Ministry of Innovation under project P08-TIC-3572.

References

1. Augusto, J.: Temporal reasoning for decision support in medicine. Artificial Intelligence in Medicine 33, 1–24 (2005)
2. Spyropoulos, C.: Ai planning and scheduling in the medical hospital environment. Artificial Intelligence in Medicine 20(101-111) (2000)
3. Votruba, P., Seyfang, A., Paesold, M., Miksch, S.: Environment-driven skeletal plan execution for the medical domain. In: European Conference on Artificial Intelligence (ECAI 2006), pp. 847–848 (2006)
4. Abidi, S.S.R.: Knowledge management in healthcare: towards 'knowledge-driven' decision-support services. International Journal of Medical Informatics 63, 5–18 (2001)
5. Coiera, E.: Clinical Decision Support Systems. In: Guide to Health Informatics (2003)
6. Peleg, M., Tu, S., Bury, J., Ciccarese, P., Fox, J., Greenes, R.A.: Comparing computer-interpretable guideline models: A case-study approach. J. Am. Med. Inform. Assoc. 10, 10–58 (2003)
7. Leong, T., Kaiser, K., Miksch, S.: Free and open source enabling technologies for patient-centric, guideline-based clinical decision support: A survey. Methods of Information in Medicine 46, 74–86 (2007)
8. Mulyar, N., van der Aalst, W.M., Peleg, M.: A pattern-based analysis of clinical computer-interpretable guideline modeling languages. JAMIA (2007)
9. Duftschmid, G., Miksch, S., Gall, W.: Verification of temporal scheduling constraints in clinical practice guidelines. In: Artificial Intelligence in Medicine (2002)
10. Terenziani, P., Anselma, L., Montani, S., Bottrighi, A.: Towards a comprehensive treatment of repetitions, periodicity and temporal constraints in clinical guidelines. In: Artificial Intelligence in Medicine, vol. 38(2), pp. 171–195 (2006)

11. Bradbrook, K., Winstanley, G., Glasspool, D., Fox, J., Griffiths, R.: Ai planning technology as a component of computerised clinical practice guidelines. In: Miksch, S., Hunter, J., Keravnou, E.T. (eds.) AIME 2005. LNCS, vol. 3581, pp. 171–180. Springer, Heidelberg (2005)

12. Ghallab, M., Nau, D., Traverso, P.: Automated Planning. Theory and Practice. Morgan Kaufmann, San Francisco (2004)

13. Castillo, L., Fdez-Olivares, J., García-Pérez, O., Palao, F.: Efficiently handling temporal knowledge in an HTN planner. In: Proceeding of ICAPS 2006, pp. 63–72 (2006)

14. Nau, D., Muoz-Avila, H., Cao, Y., Lotem, A., Mitchel, S.: Total-order planning with partially ordered subtask. In: Proceedings of the IJCAI 2001 (2001)

15. Fdez-Olivares, J., Castillo, L., García-Pérez, O., Palao, F.: Bringing users and planning technology together. Experiences in SIADEX. In: Proceedings ICAPS 2006, pp. 11–20 (2006)

16. Bresina, J.L., Jonsson, A.K., Morris, P., Rajan, K.: Activity planning for the mars exploration rovers. In: Proceedings of the ICAPS 2005, pp. 40–49 (2005)

17. Group, S.W.: National protocol for diagnosis and treatment of hodgkin's desease (in spanish).eh-seop. 2003. Technical report, SEOP (2003)

18. Fox, M., Long, D.: PDDL2-1: an extension to PDDL for expressing temporal planning domains. Technical report, University of Durham, UK (2001)

19. Allen, J.: Maintaining knowledge about temporal intervals. Comm. ACM 26(1), 832–843 (1983)

20. Shahar, Y., Young, O., Shalom, E., Galperin, M., Mayaffit, A., Moskovitch, R., Hessing, A.: A framework for a distributed, hybrid, multiple-ontology clinical-guideline library, and automated guideline-support tools. Journal of Biomedical Informatics 37(5), 325–344 (2004)

21. German, E., Leibowitz, A., Shahar, Y.: An architecture for linking medical decision-support applications to clinical databases and its evaluation. Journal of Biomedical Informatics (2008) (in press) (corrected proof)

Modeling Clinical Protocols Using Semantic MediaWiki: The Case of the Oncocure Project

Claudio Eccher[1], Antonella Ferro[2], Andreas Seyfang[3], Marco Rospocher[1], and Silvia Miksch[3]

[1] Fondazione Bruno Kessler, Trento, Italy
[2] Medical Oncology, S. Chiara Hospital, Trento, Italy
[3] Danube University Krems, Austria

Abstract. A computerized Decision Support Systems (DSS) can improve the adherence of the clinicians to clinical guidelines and protocols. The building of a prescriptive DSS based on breast cancer treatment protocols and its integration with a legacy Electronic Patient Record is the aim of the Oncocure project. An important task of this project is the encoding of the protocols in computer-executable form — a task that requires the collaboration of physicians and computer scientists in a distributed environment. In this paper, we describe our project and how semantic wiki technology was used for the encoding task. Semantic wiki technology features great flexibility, allowing to mix unstructured information and semantic annotations, and to automatically generate the final model with minimal adaptation cost. These features render semantic wikis natural candidates for small to medium scale modeling tasks, where the adaptation and training effort of bigger systems cannot be justified. This approach is not constrained to a specific protocol modeling language, but can be used as a collaborative tool for other languages. When implemented, our DSS is expected to reduce the cost of care while improving the adherence to the guideline and the quality of the documentation.

1   Introduction

The unprecedented growth in the scientific understanding and management of diseases poses the serious problem of applying this knowledge in the clinical practice. Clinical protocols adapt the available knowledge in books, articles, and clinical guidelines to the local resources and conventions at a specific site. They are a means to improve the quality of and reduce the undesired variations in care by efficiently disseminating the existing knowledge about the state of art. While they are more concise than clinical guidelines, they still can easily be constituted by several tens of pages and handling them in paper form in daily practice can be tedious. Practitioners compliance with clinical protocols and outcomes can be promoted and improved by computerized Decision Support Systems (DSS) supporting guideline-based or protocol-based care in an automated fashion at the time and location of decision making, especially when used in combination with an Electronic Patient Record (EPR) and integrated in the clinical workflow [1,2]. To this end, the Oncocure project aims at designing and implementing a prescriptive guideline-based DSS integrated with a legacy Oncological EPR (OEPR)

D. Riaño (Ed.): K4HelP 2008, LNAI 5626, pp. 42–54, 2009.

in use in the Medical Oncology Unit (MOU) of the S. Chiara Hospital of Trento (Northern Italy). The DSS is based on the Asbru [3] encoding of protocols of breast cancer medical therapies used in the unit.

The translation process when dealing with protocols shares important characteristics with the modeling of clinical guidelines. The difference is that the protocols are much more concise than guidelines and their structure is nearer to a formal representation. Therefore, the transformation effort is smaller for protocols. Nonetheless, it is far from trivial, since protocols, like guidelines, are combinations of written text and informal diagrams, and contain implicit knowledge and assumptions about the care process and the medical background, which need to be acquired during the modeling process. It requires the collaboration of physicians and computer scientists: only the former have the knowledge to grasp the deep clinical meaning integrated in the protocols, and only the latter have the training in creating formal models. Moreover, members of the encoding team may not be located in the same building, or even in the same town, and may not be able to physically participate in meetings.

In the case of the Oncocure project, the modeling phase required collaboration between oncologists and computer scientists located in Trento and in Vienna. Interaction had to be supported by a continuous exchange of text and pseudo-code based documents, which had to be maintained, updated and all changes had to be traced. Thus, we needed a web-based tool for intuitive collaborative editing to elaborate the model and exchange notes. In this paper, we present the Oncocure project and describe the use of Semantic MediaWiki (SMW) as a collaborative framework for the knowledge acquisition phase of the project. The proposed approach allows remotely located people to actively participate to the encoding of the breast cancer protocols into the skeletal plan-representation language Asbru.

In Section 2 we describe the Oncocure project. In Section 3 and 4 we present the modeling phase and the process of protocols encoding using SMW, by showing a running example. Finally, in Section 5 we discuss the related work and draw some conclusions.

2   The Oncocure Project

Clinical Practice. Inside the MOU of the S. Chiara Hospital of Trento, each oncologist is specialist in one or more type of cancer (e.g., breast cancer, colon-rectal cancer, etc.) However, in the daily routine at the hospital, he/she is often required to treat patients with other types of cancer. Medical treatment to cancer patients is also provided in the Internal Medicine wards of the peripheral hospitals of the Province of Trento, which lack a specific oncology service, under the supervision of an oncologist of the MOU. In such cases, the DSS will be particularly useful, as it may efficiently assist the physician in looking up the required knowledge.

Central to the care process are the periodical encounters with the patients, in which the physician visits the patient and decides the appropriate strategy on the basis of an objective examination and laboratory and radiological reports: whether to continue, suspend or interrupt an ongoing medical therapy or initiate a new one.

Infrastructure. Since 2000, a web-based OEPR, developed in our laboratory [4], has been in use in five hospitals of our Province. OEPR provides a first level of "passive" support to the shared management of cancer patients between the Medical Oncology Unit of the S. Chiara Hospital and the peripheral hospitals, allowing to store patient and cancer data as well as past and ongoing therapies and outcomes. Currently, all the cancer patients are managed through the system, which up to now stores more than 10 000 cases.

On this legacy infrastructure, we intend to add intelligent tools for actively supporting the physicians in their everyday clinical practice. To this end, the two year *Oncocure* project started in April 2007, with the main aim to design and develop a prescriptive guideline-based DSS for giving active support at important decisional steps of the oncological care process. The DSS is based on the Asbru language, used to encode the protocols of pharmacological therapies for breast cancer in use in the MOU of the S. Chiara Hospital, and integrates the Asbru interpreter [5]. The system will generate patient-specific reminders about the most appropriate therapeutic strategy recommended by the cancer treatment protocols in the presence of the specific disease and patient conditions. To this end, the DSS must be seamlessly integrated with the legacy OEPR and the local clinical workflow. At the same time, we wanted an architectural solution ensuring loose coupling between the DSS and the legacy OEPR, in order to build an easy maintenance system integrable with different clinical information systems. For this reason, the DSS will be built as a Web service invokable by the physician from the OEPR User Interface (see Figure 1). Moreover, modifying the knowledge representation or the core of the existing execution engine for the specific database structure is not a feasible option

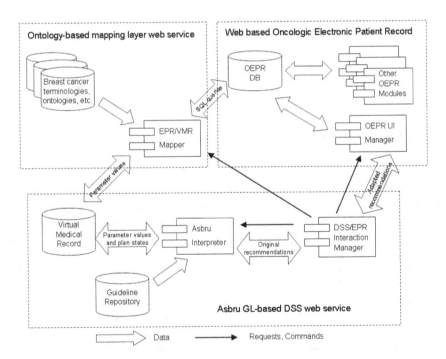

Fig. 1. Block diagram of the DSS integrated with the legacy OEPR. The VMR and ontology-based mapping layer separates the DSS from the OEPR.

from the software engineering perspective. Like other teams in comparable situations, therefore, we utilize the Virtual Medical Record (VMR) approach [6], which supports a well-defined structured data model for representing information related to individual patients. A breast cancer VMR will be defined, in which the parameters required by the DSS, extracted or abstracted from the OEPR database through an ontology-based mapping layer, are stored along with the states reached by the interpreter during guideline execution. Ontologies, in fact, are widely acknowledged as the essential glue to ensure semantic consistency to data and knowledge [7]. This approach allows to deploy our solution at other points of care.

Aim and Evaluation. An important part of the work is the test phase, planned for the last part of the project, which will give indications about the real effectiveness and the impact on the clinical practice of a protocol-based DSS integrated into the careflow. According to the oncologists, we expect several benefits from the deployment of the Oncocure system, especially for those oncologists of the MOU who do not specifically follow the breast cancer disease and for the clinicians of the peripheral hospitals who provide breast cancer medical treatments.

– Improved adherence to protocols, because all the decisions can be taken after consultation of the protocols. This, in turn, can improve the care quality by favoring the use of the best-evidence.
– Reduction of labor of the clinicians, because for each case only the relevant protocol fragments are displayed. This avoids the necessity for physician to leaf through protocols to find the right recommendation, especially in the presence of the patient.
– Improved documentation, because each decision of the physician can be automatically logged in the OEPR. If a valid justification is required from the care provider who refuses a recommended treatment, this information can be used to verify *a posteriori* the actual applicability of the protocols and to improve their quality.

3   Modeling Background

The internal breast cancer treatment protocols are a local customization of national guidelines issued by AIOM (Italian Association of Medical Oncology) [8] integrated with S. Gallen conference recommendations [9], prepared by the breast cancer specialist of the MOU. In contrast to guidelines, which are mostly in textual form, the internal protocols are mainly constituted by informal "box and arrow" diagrams accompanied by short explanation text. In Figure 2 we show an example of the diagram for the adjuvant therapy decision of hormone responsive patients.

Notwithstanding protocols are more formal and concise than guidelines, they still present ambiguities and incompletenesses that must be resolved. Moreover, to smoothly integrate the system in the clinical workflow, it is necessary to explicit the care process tasks in which protocols are used and define the system requirements. Consequently, the knowledge acquisition phase was still considered the most important task of the Oncocure project. Most of the first year of the project (10 months), in fact, has been devoted to the acquisition of the explicit knowledge written in the protocols and the implicit knowledge on the real care process carried out in the ward, and to the codification

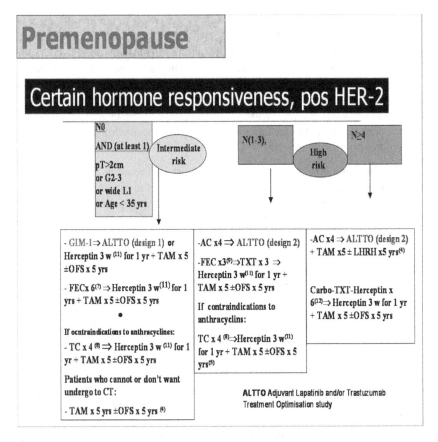

Fig. 2. An example of the informal diagrams of the adjuvant treatment decision in the breast cancer internal protocols prepared by the oncologist. It is quite concise, but a certain amount of knowledge is not explicit (e.g., the preferences/conditions for the choice of one of several therapies.)

of the Asbru model. This knowledge acquisition phase was conducted through regular meetings between a computer scientists and an expert oncologists in Trento, with the intervention of a computer scientists in Vienna by phone or through the preparation of documents, email, etc., discussed during the encounters.

Although the size of the protocol and the resulting Asbru model is not large, we still needed a multi-user, web-based tool to manage all the pieces of knowledge which accumulated over time, and to ensure that everyone is working on the same updated version. Semantic MediaWiki [10] belongs to a new generation of tools supporting the integration of Web 2.0 and Semantic Web approaches [11] that have been developed to meet the needs of the semantic modeling community to easily create, share, and connect content and knowledge. While preserving the freedom in format of the overall page, it introduces knowledge annotations with a formal semantics for their use. This is an advantage over editors like Uruz [12] which enforce the structure of Asbru plans onto the data entry forms. Moreover, all the important functionalities (access control and permissions, tracing of the activity, semantic search, etc.) are already provided by the SMW framework, without needing to install specific client applications.

SMW employs binary relations and attributes to render machine-processable the shared knowledge of wiki. Key semantic annotations of SMW are the following:

- *Categories*, which classify the pages according to their content.
- *Typed links*, which express a relation between pages. New types can be created by the user on the fly by just employing them for annotations.
- *Attributes*, which specify simple datatype properties related to the content of a page.

The access to the wiki pages is protected by login and password. Different level of permissions can be defined for users.

The SMW framework provides RDF exports of the pages, which can be transformed into other representations. This means that, while the overall wiki page can be organized as the authors prefer, the collection of semi-formal fragments can be automatically transformed into a basis for the final Asbru model.

One of the disadvantages of SMW is the requirement to handle the special syntax. This is generally not perceived to be easy by non-IT people involved in the authoring process.

4   The Encoding Process

The basic idea is that an Asbru guideline model is expressed as a collection of inter-related SMW pages connected by typed links. A SMW page corresponds to an Asbru building block of the guideline and may contain:

1. typed links connecting the page to other building block pages (e.g., a plan and its subplans),
2. attributes for expressing element data that must be present in the model (e.g., the title of a plan), and
3. free text, organized in sections, for documenting the model and clarifying it to users not trained in the formal representation (e.g., reference to source documents, annotations about modeling choices and open problems, etc.) The comments can be added by each user and are not translated in the final model.

Each page is assigned to one of a set of predefined SMW categories according to the type of Asbru element it models, by means of a specific semantic annotation. In Figure 3 we show the diagram of the process steps for encoding the protocols in Asbru using SMW.

Creation of the SMW Pages. To facilitate the creation of wiki pages, we have defined a set of template pages for the most frequently used Asbru building blocks. For this first version of the tool, we have defined a template for each plan body type (*plan using subplans, plan activation, ask, user performed, cyclical plan, variable assignment, if-then-else*), two templates for parameters (*raw data definition* and *qualitative parameter definition*), and two templates for abstraction (*qualitative scale definition* and *secondary qualitative entry.*)

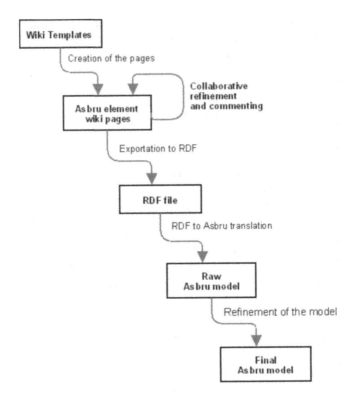

Fig. 3. Diagram of the process to encode the protocols in Asbru using SMW. The activity in bold is the phase in which model pages are collaboratively revisioned and consolidated.

A new Asbru element can be defined simply by cutting and pasting the template into the editing area of the new wiki page, completing the fields and removing those not required (e.g., unnecessary conditions in plans). However, to overcome the drawback of handling SMW syntax of slots, we designed a Java page generator applet with a user interface, shown in Figure 4, that presents fill-in fields the user can complete to define the page content. The applet allows the user to choose one of the predefined page templates and automatically generates the interface layout from the chosen template, so that the layout can be adapted to changing requirements during the project with minimal effort. Of course, this takes some freedom in the organization of the content from the users, but it relieves them from learning the syntax and it helps to prevent editing errors. Also, the use of the page generator does not prevent later individual changes to pages, without any restraints.

Once the required fields have been completed, the wiki page can be saved and uploaded into the SMW framework. In Figure 5 we show the wiki page generated from the *plan using subplans* source template, which models the set of chemotherapy treatments recommended in the middle box of Figure 2. The corresponding wiki code with the semantic annotations in double square brackets is displayed in Figure 6.

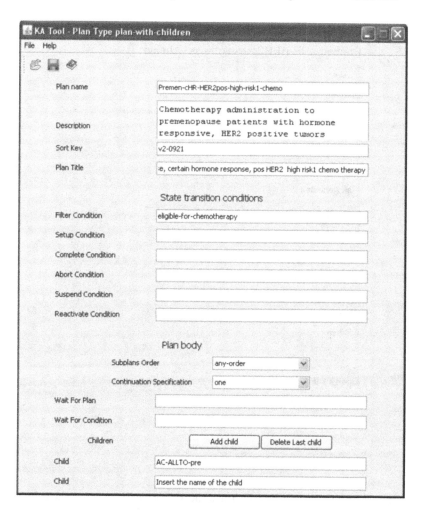

Fig. 4. User interface of the Java tool for loading and editing the templates to create a SMW page. The example shows the layout for a *plan using subplans* template. The user can fill in the fields adding the required information and, for this particular plan, add as many children as needed. Then the page is saved and directly uploaded in wiki (see next figure).

An important knowledge slot is the reference to the exact location in the source document, on which the model fragment described by a particular wiki page is based. This was inspired by MHB and the DELT/A tool [13], where links are inserted automatically. In Oncocure, the source document is technically difficult to access (graphics in a non-editable data format). Therefore, we resorted to manually inserting a code with page number and a separate order key within the page (Sort key in Figure 5) in the wiki model. Using both, we can print the wiki model in the precise order of the source document, which allows the easy side-by-side comparison of source and model also by non-IT people.

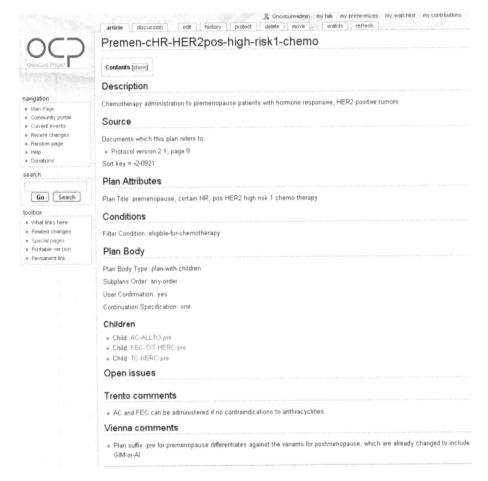

Fig. 5. Snapshot of the SMW page of the plan modeling the chemotherapy treatment options in the middle box of Figure 2. In the *Open issues*, *Vienna comments* and *Trento comments* sections each participant can add comments and explanations. Note the *Sort key* entry, which allows to relate the wiki page (and the generated Asbru plan) to the precise position in the source document.

Page Refinement and RDF Export. Pages can be refined and commented by each member of the modeling team. Once the team has agreed upon the general modeling issues, a built-in functionality allows to choose a set of pages and export them in a RDF file. Categories, pages and semantic annotations are exported as classes, instances, and properties, respectively.

Asbru Model Generation and Refinement. A second custom Java applet transforms the RDF export to an Asbru XML model according to the Asbru DTD (see Figure 7). This model has those fields set to their final value, for which sufficient and precise information is available. For those cases, where the SMW model contains only a sketch in free text (e.g., the conditions on plan state transitions in this first version), the available information is inserted as a comment and the refinement is left for the next modeling

```
== Description ==

[has Description:=Chemotherapy administration to premenopause patients with
hormone responsive, HER2 positive tumors]]

== Source ==

Documents which this plan refers to:

* Protocol version 2.1, page 9

Sort key = [[has Sort Key:=v2-0921]]

== Plan Attributes ==

Plan Title: [[has Plan Title:=premenopause, certain hormone response, pos HER2
high risk1 chemo therapy]]

==Conditions==

Filter Condition: [[has Filter Condition:=eligible-for-chemotherapy]]

== Plan Body ==

Plan Body Type: [[has Plan Body Type:=plan-with-children]]

Subplans Order: [[has Subplans Order:=any-order]]

User Confirmation: [[has User Confirmation:=yes]]

Continuation Specification: [[has Continuation Specification:=one]]

=== Children ===

* Child: [[has Child::AC-ALLTO-pre]]
* Child: [[has Child::FEC-TXT-HERC-pre]]
* Child: [[has Child::TC-HERC-pre]]
```

Fig. 6. SMW code defining the plan in Figure 5. Only semantic annotations, included between double square brackets, are translated in RDF and then in Asbru. For example, the current plan is defined as a *plan using subplans* ([[has Plan Body Type:=plan-with-children]]) with three children (AC-ALLTO-pre, FEC-TXT-HERC-pre, and TC-HERC-pre).

phase. The Asbru model is then further refined and completed (e.g., transforming the pseudo code conditions into Asbru propositions) through a dedicated Asbru modeling software like DELT/A.

5   Related Work

A number of Clinical Practice Guidelines (CPG) frameworks representing task, plan and decision structures have been proposed in recent years. Some of them have been complemented with active software tools for guideline execution [14]. See Peleg et al. [15] for a detailed comparison and Mulyar et al. [16] for a pattern-based analysis of CPGs. In both comparisons, Asbru [3] scored favorably.

Guideline Modeling Tools can be roughly classified into (1) model-centric and (2) document-centric approaches. In the model-centric approach, a conceptual model is formulated by domain experts. Thus, the relationship between the model and the original document of the clinical guideline is only indirect. In the document-centric approach

```
<plan name="Premen-cHR-HER2pos-high-risk1-chemo" title="premenopause, certain
            hormone response, pos HER2 high risk1 chemo therapy">
  <administrative-data>
    <comment text="v2-0921" />
  </administrative-data>
  <comment text="Chemotherapy administration to premenopause patients with
                 hormone responsive, HER2 positive tumors"/>
  <conditions>
    <filter-precondition confirmation-required="yes" overridable="no">
      <comment text="eligible-for-chemotherapy" />
      <to-be-defined />
    </filter-precondition>
  </conditions>
  <plan-body>
    <subplans retry-aborted-subplans="no" type="any-order"
               wait-for-optional-subplans="no">
      <wait-for>
        <one/>
      </wait-for>
      <plan-activation>
        <plan-schema name="AC-ALLTO-pre"/>
      </plan-activation>
      <plan-activation>
        <plan-schema name="FEC-TXT-HERC-pre"/>
      </plan-activation>
      <plan-activation>
        <plan-schema name="TC-HERC-pre"/>
      </plan-activation>
    </subplans>
  </plan-body>
</plan>
```

Fig. 7. Fragment of the valid skeletal Asbru model generated from the SMW semantic annotations shown in Figure 6. The Asbru model is the translation of the RDF dump generated using the SMW built-in function.

markup-based tools are used to systematically mark up the original guideline in order to generate a semi-formal model of the marked text part (see Leong et al. [17] for an overview.) In our project, both approaches merged to a considerable extent, as the protocol is already in a structure similar to a formal model. One could say that we followed a model-centric view, without loosing contact with the document structure, by keeping precise references to the source document. In a previous project [18], an intermediate notation, MHB [13], was used to model a breast cancer guideline. The protocol modeled in our current project, however, does not contain as many dimensions of knowledge as the previously modeled guideline. Also, the smaller bridging effort and the tense time frame of the project suggested going from the original and already well-structured document to Asbru, without an intermediate representation. Still, we needed a multi-user, web-based tool that had to allow a significant amount of freedom of the form, which is one of the prime advantages of wikis.

6   Conclusion

In this paper, we present the Oncocure project and propose to use SMW as a collaborative tool for knowledge acquisition of treatment protocols. To our knowledge there is only a paper in biomedical informatics describing the use of an extended

SMW (BOWiki) for the collaborative annotation of gene information [19]. Main advantages of the solution we propose are:

1. the provision of a lightweight infrastructure that allows each participant to easily follow and contribute to the encoding process,
2. the possibility to mix informal content with semi-formal fragments, and
3. the automatic generation of valid skeletal Asbru models.

The oncologist expressed a positive attitude toward this collaborative modeling tool, because she was able to easily understand and verify the computer scientists's modeling work, using both the page structure and the informal annotations and comments. SMW, however, still requires the knowledge of the Asbru language to correctly complete the semantic tags; it does not solve the problem of allowing a domain expert to work alone in encoding guidelines to maintain/update the knowledge base. However, this is a problem shared by all knowledge acquisition projects. Its solution requires a broader approach than just applying software, e.g., user training in formal modeling. Although our work is focused on using Asbru as encoding language, our approach can be generalized to encode protocols in other languages in a collaborative way, by defining templates with slots specific for the language used.

This first prototype is limited to the most frequently used Asbru elements. Further work is needed for the definition of templates for more complex blocks. We are also working to fully integrate into the SMW framework the page generator interface and the conversion from RDF to Asbru, so as to generate wiki pages and Asbru code from within the pages themselves, and to provide a built-in navigable graphical representation of the model.

Acknowledgment

This work has been carried out in the context of the project "*Oncocure - Information and Communication Technologies for clinical governance in oncology: Design and development of a computerized guideline-based system for supporting and evaluating evidence-based clinical practice*", funded by the Fondazione Caritro of Trento.

References

1. Sonnenberg, F., Hagerty, C.: Computer interpretable guidelines: where are we and where are we going? 2006 IMIA Yearbook of Medical Informatics, Methods Inf. Med. 45(suppl.1), S145–S158 (2006)
2. Bates, D., Kuperman, G., Wang, S., Gandhi, T., Kittler, A., Volk, L., Spurr, C., Khorasani, R., Tanasi-jevic, M., Middleton, B.: Ten commandments for effective clinical decision support: making the practice of evidence-based medicine a reality. J. Am. Med. Inform. Assoc. 10, 523–530 (2003)
3. Seyfang, A., Kosara, R., Miksch, S.: Asbru 7.3 reference manual. Technical report, Vienna University of Technology (2002)
4. Eccher, C., Berloffa, F., Galligioni, E., Larcher, B., Forti, S.: Experience in designing and evaluating a teleconsultation system supporting shared care of oncological patients. In: Proceedings of AMIA Symposium 2003, p. 835 (2003)

5. Seyfang, A., Paesold, M., Votruba, P., Miksch, S.: Improving the execution of clinical guidelines and temporal data abstraction in high-frequency domains. In: tenTeije, A., Lucas, P., Miksch, S. (eds.) Computer-Based Medical Guidelines and Protocols: A Primer and Current Trends. Health Technology and Informatics. IOS Press, Amsterdam (2008) (forthcoming)
6. Johnson, P., Tu, S., Musen, M., Purves, I.: A virtual medical record for guideline-based decision support. In: Proceedings of AMIA Symposium 2001, pp. 294–298 (2001)
7. Pisanelli, D., Battaglia, M., De Lazzari, C.: ROME: a reference ontology in medicine. In: Pisanelli, D., Fujita, H. (eds.) New Trends in Software Methodologies, Tools and Techniques. Frontiers in Artificial Intelligence and Applications, pp. 485–493. IOS Press, Amsterdam (2007)
8. http://www.aiom.it/it/Oncologiamedica/Lineeguida.asp
9. Harbeck, N., Jakesz, R.: St. Gallen 2007: Breast Cancer Treatment Consensus Report. Breast Care 2, 130–134 (2007)
10. Völkel, M., Krötzsch, M., Vrandecic, D., Haller, H., Studer, R.: Semantic Wikipedia. In: WWW 2006, Edinburgh, Scotland, May 23-26 (2006)
11. Noy, N., Chugh, A., Alani, H.: The CKC challenge: exploring tools for collaborative knowledge construction. IEEE Intelligent Systems 23(1), 64–68 (2008)
12. Shahar, T., Young, O., Shalom, E., Galperin, M., Mayaffit, A., Moskovitch, R., Hessing, A.: A framework for a distributed, hybrid, multiple-ontology clinical-guideline library, and automated guideline-support tools. Journal of Biomedical Informatics 37, 325–344 (2004)
13. Seyfang, A., Miksch, S., Marcos, M., Wittenberg, J., Polo-Conde, C., Rosenbrand, K.: Bridging the gap between informal and formal guideline representations. In: Brewka, G., Coradeschi, S., Perini, A., Traverso, P. (eds.) 17th European Conferencen on Artificial Intelligence, pp. 447–451. IOS Press, Amsterdam (2006)
14. de Clercq, P., Blom, J., Korsten, H., Hasman, A.: Approaches for creating computer-interpretable guidelines that facilitate decision support. Artif. Intell. Med. 31, 1–27 (2004)
15. Peleg, M., Tu, S., Bury, J., Ciccarese, P., Fox, J., Greenes, R., Hall, R., Johnson, P., Jones, N., Kumar, A., Miksch, S., Quaglini, S., Seyfang, A., Shortliffe, E., Stefanelli, M.: Comparing computer-interpretable guideline models: a case study ap-proach. J. Am. Med. Inf. Assoc. 10(1), 52–68 (2003)
16. Mulyar, N., van der Aalst, W., Peleg, M.: A pattern-based analysis of clinical computer-interpretable guideline modeling languages. J. Am. Med. Inf. Assoc. 14(6) (to appear)
17. Leong, T., Kaiser, K., Miksch, S.: Free and open source enabling technologies for patient-centric, guideline-based clinical decision support: a survey. 2007 IMIA Yearbook of Medical Informatics, Methods Inf. Med. 46(1), 74–86 (2007)
18. ten Teije, A., Marcos, M., Balser, M., van Croonenborg, J., Duelli, C., van Harmelen, F., Lucas, P., Miksch, S., Reif, W., Rosenbrand, K., Seyfang, A.: Improving medical protocols by formal methods. Artif. Intell. Med. 36(3), 193–209 (2006)
19. Backaus, M., Kelso, J.: BOWiki - a collaborative annotation and ontology curation framework. In: WWW 2007, Banff, Canada, May 8-12 (2007)

A Conceptual Framework for Ontology Based Automating and Merging of Clinical Pathways of Comorbidities

Samina Raza Abidi

NICHE Research Group, Faculty of Computer Science, Dalhousie University, Halifax, Canada

Abstract. In this paper we present a conceptual framework for ontology based knowledge representation and merging of Clinical Pathways (CP) of comorbidities. Paper based CP are static documents which do not have adaptability to accommodate dynamic changes in a patients conditions, particularly in case of co-morbidity, since most CP are focused on a single disease management. Our approach to computerize and merge CP of comorbidities for decision support purpose include; 1. Representation of comorbidity CP as OWL ontologies, 2. Merging ontologies along common tasks, 3. Execution of merged ontology using OWL reasoner to provide CP mediated decision support, 4. Evaluation of recommendations by instantiating the ontology with comorbidity scenarios. Most challenging and unique aspect of this research is that it involves the dynamic integration of computerized CP of two concurrent comorbid diseases, whilst maintaining clinical pragmatics and medical correctness. We believe that Semantic Web has enormous potential to achieve this goal.

1 Introduction

Clinical pathways (CP) are structured tools which formalize various pre-determined steps in the health care processes and make them predictable. Being static documents, they do not have flexibility or adaptability to accommodate dynamic changes in a patients conditions. This is particularly true if any comorbidity exist in a particular individual. A CP is virtually useless in such circumstances [1]. Comorbidities are regarded as conditions that exist at the same time as the primary condition in the same patient. According to a 2004 Canadian study by the national center of health statistics about 10% of the population with ages 60 years or younger have at least three chronic diseases, while people over 60 years of age have at least seven concurrent illnesses. Numerous published studies have reported health and economic burden of chronic diseases with comorbidities. Utilization of the hospital resources, physicians services, and length of hospital stay increases exponentially as the number of comorbid conditions increases. Most available CP however, are focused on a single disease management and does not take into account individual clinical features [2]. The term "adaptable clinical pathway" is used to signify step-wise automation of CP for each patient in accordance to his/her health profile. In this paper we present a conceptual framework for dynamic integration of computerized CP of two concurrent comorbid diseases whilst preserving clinical pragmatics and medical precision. This research aim to translate evidence-based

D. Riaño (Ed.): K4HelP 2008, LNAI 5626, pp. 55–66, 2009.

knowledge contained in CP into clinical decision making and care planning processes for management of chronic diseases and their comorbidities. We note that the health knowledge management literature constitutes a number of clinical guideline representation formalisms, such as GLIF [3], EON [4], SAGE [5] and Proforma [6]. Yet, the problem of CP merging has not yet been adequately addressed in the literature.

2   Problem Description and Rationale

In order to achieve desired CP adaptability, our challenge is to;

1. Automate CP without jeopardizing the underlying semantics,
2. Merge CP along common tasks or actions while maintaining the integrity and consistency of the medical knowledge,
3. Abstract clinically significant rules from the innate decision logic in the pathway document,
4. Execute the automated CP to infer recommendations.

We believe that the above challenges can be met by adopting Semantic Web approach, which allows explicit representation of CP knowledge in the form of computer interpretable Web Ontology Language (OWL) ontologies. These ontologies can then be integrated to achieve a merged comorbidity pathway. Since ontology formalizes the inherent decision logic in the pathway documents, it can be reasoned about by computers with rules, thus providing decision support. We believe that using OWL as a representation formalism has some advantages over other existing guideline formalisms. OWL ontology allows flexibility of encoding domain knowledge as declarative model, depicting classes, their relations and constraints of these relations, which are interpretable by both humans and intelligent applications. In addition, being derived from descriptive logic, OWL supports formal semantics and consequently reasoning support. Thus OWL based reasoning applications can benefit from high performance reasoners such as Racer Pro.

In a related research we have successfully implemented a Breast Cancer Follow-up Decision Support System based on a Breast Cancer Follow-up Clinical Practice Guideline (CPG) [7]. We employed a Semantic Web approach to model the CPG knowledge and to reason over the CPG derived OWL ontology to provide trusted CPG-driven recommendations. However there are some fundamental structural differences between the two entities; development and scope of implementation of CP, unlike CPG is multidisciplinary. While a CPG provides evidence based recommendations regarding diagnosis or therapy etc., CP focuses on improving efficiency and quality of care once these decisions have been made. Thus CP in contrast to CPG are designed along precise timelines so that sequencing and scheduling of the tasks are laid out explicitly. CP contain intermediate outcomes which serve as indicators of pathway performance and progress of patient. In addition a CP allows documentation of any deviation from the expected course in the form of variance [2]. Therefore, given these differences between CPG and CP, the modeling and execution of a CP poses different challenges than a CPG.

In this paper we present an approach to computerize and merge CP of comorbidities. We believe that Semantic Web offers logic based framework to model and merge the

CP, whilst maintaining clinical pragmatics and medical correctness. Our methodology include four phases; (1). Representation of CP as OWL ontologies, (2). Merging CP by identifying similar semantic patterns in source ontologies, (3). Execution of the merged ontology using logic based OWL reasoner to achieve CP mediated decision support, (4). Evaluation of recommendations after instantiating the merged ontology with clinical scenarios involving comorbidities.

3   Computerization of Comorbidity Pathway: An Ontological Approach

Clinical decision support based on clinical knowledge in the pathway documents is an extremely complex endeavor since it involves dynamic interaction between various clinical knowledge components. These include patient's specific clinical characteristics which are dynamic and evolve over time, operational or procedural constraints specified in the pathway document, and patient's progress paths with desired outcomes. Since a clinical decision support system is only as effective as its underlying knowledge base, it is imperative to develop a sound methodology to develop a high quality knowledge base.

Our methodology to design CP based ontologies has been adapted from Knowledge Acquisition Life Cycle (KAC), an approach to build knowledge based systems [8]. This model captures important aspects of knowledge acquisition and modeling process. CP based decision support require modeling of both declarative and procedural knowledge inherent in clinical pathways. We believe our ontology based methodology is sound and comprehensive enough to explicitly articulate facts about categorization of the relevant concepts and their relationship in the domain and the manner in which this declarative knowledge can be inferred on.

As shown in Figure 1, our approach to ontology design consist of four steps; Identification, Conceptualization, Formalization/Implementation and Testing of Knowledge. These steps are executed recursively in an iterative refinement process till a representative, valid knowledge base is achieved.

3.1   Identification and Classification of CP Knowledge

We have identified and classified CP knowledge as;

1. Medical knowledge i.e. knowledge regarding primitives such as, actions or tasks and decision criteria
2. Process knowledge such as, available resources and scheduling constraints
3. Patient knowledge, such as clinical features and demographic information
4. Proof rules which are constraints either explicitly stated in the CP or implicitly derived while merging the pathways

The identification and classification of CP knowledge is carried out with the help of domain experts, and available medical literature. In addition, having been trained as a physician herself, the author also utilized her own tacit knowledge for this task.

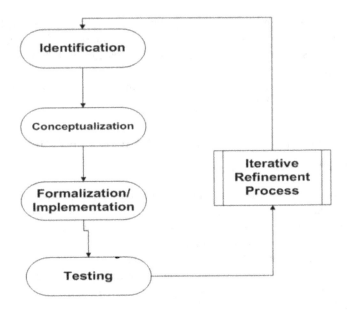

Fig. 1. Knowledge modeling approach

3.2   Conceptualization of Identified Knowledge

Three most significant knowledge components associated with conceptualization (Figure 2) of textual language include [9];

1. Syntactic Knowledge; which is used to determine the structure of a sentence. As an example, consider this statement, "only a registered nurse can titrate Beta Blocker". In this case "registered nurse" is an object (noun), "can titrate" is predicate (verb) and "beta blocker" is subject (noun).
2. Semantic Knowledge; which is used to determine the meaning of the words and the way they are amalgamated to form a consequential sentence. For example; "Patient presented with history of ongoing dyspnea, aggravated with exertion with no angina". This means that patient had history of shortness of breath, the shortness of breath worsens with any physical exertion and there was no chest pain (angina).
3. Domain Knowledge; which contains the information regarding the specific subject matter. For example Beta Blocker is a medication or dyspnea is shortness of breath.

Designing Ontology necessitate achieving semantically unambiguous coverage of relevant medical domain as much as possible. This means, not only to capture the domain knowledge but also to provide an explicit and agreed upon understanding of the domain in terms of related syntactic and semantic knowledge. Therefore in this phase primary concepts in the CP knowledge and their relationship and constraints on the relationships are explicitly stated.

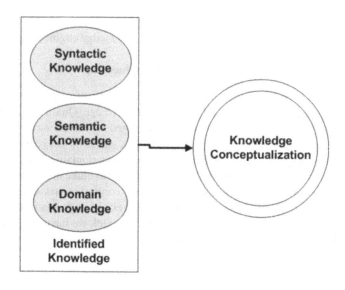

Fig. 2. Knowledge components associated with conceptualization

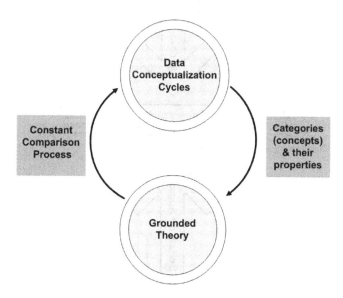

Fig. 3. Knowledge conceptualization based on Grounded Theory

Since conceptualization is core category in Grounded Theory (GT) [10], we have adopted the fundamentals of GT to direct this phase in our methodology (Figure 3). GT offer constant comparison process to generate conceptualizations from data into integrated patterns, which are denoted by categories (concepts) and their properties [10].

In order to achieve reliable and accurate conceptualization our research involves multidisciplinary collaboration between the clinical and knowledge engineering disciplines.

3.3 Knowledge Formalization and Ontology Engineering

We have employed Web Ontology Language (OWL) to formally represent conceptualized CP knowledge as ontology using ontology editor Protege. Quality of a knowledge representation model such as an ontology can be judged by certain basic criteria such as stability, consistency and correctness of a model. A stable model require minimum alteration to reflect any new requirements or additions in a domain. Thus, if the ontology remain stable in such scenarios, it is less troublesome to make necessary variations. A consistent model should not have redundancies and contradictions. Consistency can be ensured by identification of commonality of interrelated concepts in domain [11]. In order to ensure the stability and consistency of the model, we have adopted "middle out" [11] approach to ontology engineering (Figure 4).

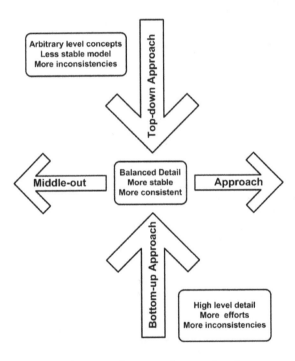

Fig. 4. Approaches to ontology engineering

This approach provides a good balance in terms of level of detail. In this approach, ontology engineering process starts with specifying most important concepts in the domain. The detail in the description of these concepts arises only as necessary. The higher level concepts in the domain are defined in terms of these concepts. As a result the higher level categories arise naturally. Such model is more likely to be stable and more consistent.

3.4 Testing Ontologies for Validity; Class Hierarchy and Logical Consistency

Ontological models are useful for proofs since they fulfill certain requirements such as consistency and completeness that automatically apply to the inference systems which use them. The CP based Ontologies are tested for class hierarchy and logical consistency which is carried out during the reasoning process. Inconsistencies in a model can be identified and highlighted with the help of an OWL compliant reasoner, which then can be corrected manually.

3.5 Specific Challenges with Regards to CP Based Ontology Design

A Clinical Pathway for a single disease management, being a multidisciplinary artifact, may consist of several categories of care and several paths for disease progression and desired outcomes. In a pathway document, for example, one for acute coronary syndrome, there would be processes such as consultation, assessment, investigation, treatment, follow-up, etc. Within these processes there are specific activities and tasks, for example meeting with cardiology coordinator nurse, performing ECG etc.(Figure 5)

	Day of Admission (ER – HI)	Day 1	Day 2	Day 3
Date (YY/MM/DD)				
Consults	• Cardiology Nsg Coordinator PRN (HI) • Smoking Cessation: ☐ Yes ☐ N/A	• Social work consult PRN ___ • Patient Revascularization Registry Form Completed: ☐ Yes ☐ No	• Cardiac Rehabilitation ___ • Patient Revascularization Registry Form Completed: ☐ Yes ☐ No	• Dietitian for ↑ Lipids & PRN ___ • Physiotherapy PRN ___ • Occupational Therapy PRN ___ • Patient Revascularization Registry Form Completed: ☐ Yes ☐ No
Tests	• ECG ___ • Chest x-ray ___ • CBC, platelets, Na, K, Cl, Cr, glucose, INR, PTT ___ • Troponin q 8 h × 2 ___ • CK q 8 h × 3 ___ • If patient a known diabetic *OR* fasting glucose results >7.0 mmol/L *OR* random glucose results >11.0 mmol/L then do capillary glucose monitoring (CBG) QID • MRSA / VRE swabs sent: ☐ Yes ☐ No ☐ NA	• ECG ___ • Consider Echo (AWMI) for day 3–4 • Fasting cholesterol, HDL, LDL, triglycerides, fasting glucose & Hba1c a.m. of Day 1	• ECG ___ • Confirm PA & Lateral Chest x-ray done	
Assessments / Treatments	• Cardiac Monitor ___ • O₂ by Titration Protocol ___ • VS q4h while awake ___ • Chest pain protocol ___	• Cardiac Monitor ___ • O₂ by Titration Protocol ___ • VS QID ___ • Chest pain protocol ___	• Ask Physician re Monitor ___ • O₂ by Titration Protocol ___ • VS BID & PRN ___ • Chest pain protocol ___	• VS BID & PRN ___ • O₂ by Titration Protocol ___ • Chest pain protocol ___

Fig. 5. A segment from Acute Coronary Syndrome Clinical Pathway showing care processes and specific tasks within them and the time frame within which these tasks are to be done

These specific tasks within different categories of care are to be performed within specific time frame. This is because most CP are designed as time-task matrix specifying time intervals in which various tasks are carried out and time interval between the tasks. In addition, activities are time and date stamped. We believe that these task dependencies and temporal constraints can be best represented by incorporating relevant concepts from OWL Time ontology in the Model. OWL Time ontology provides relationships between the classes such as, CalenderClockDescription, DurationDescription, Interval, IntervalEvent, Instant and InstantEvent. With relationships such as, End, Begin, Before and After, the classes; Interval, Instant, IntervalEvent, and InstantEvent can be used to represent the task dependencies.

All pathway documents contain variance record to document any deviation from the expected course. Variance may be related to a specific task, an outcome, or a time frame within which a task has to be completed. In order to model the variance in the pathway our challenge is three fold:

1. When the variance has taken place?
2. What is the reason of the variance?
3. What is the next action plan for the patient?

An adaptable computerized pathway must allow clinicians, flexibility of changing the plan to correlate with patients changing needs. Our approach, in this regard is to automatically identify variance when patients assessment data is evaluated against relevant progress criteria defined for each progress node in the ontology, suggesting that the required outcome has not been achieved. Possible reasons of variance and alternate action plans are modeled using our own tacit knowledge and that of domain experts and available relevant explicit knowledge.

Desired outcomes, recommendations and clinical information in the pathways of comorbidites can be personalized by creating specific scenarios pertaining to patients clinical status. The patient profiles are generated by collecting, aggregating and representing data regarding clinical characteristics such as any diagnostic test measurement, sign or symptom, or therapeutic element, as specified in the pathways. This is accomplished by creating a class "Patient" in the ontological model. The class Patient has range of properties collecting relevant data to represent a specific patient profile. This approach ensures that the pathways would be dynamically able to adapt to a specific patient profile and would generate a patient-specific care plan.

4   Merging Pathways of Comorbidities

Once the comorbidity pathways have been formalized in the form of ontologies, they can be merged by integrating two ontologies. Merging CP based ontologies involves integrating knowledge affirmed in classes, their relationships and the execution rules with respect to a number of entities such as (Figure 6) 1. Medical domain knowledge, 2. Patient data, 3. Operational data, 4. Infrastructural data. The merging of source ontologies results in creation of a single coherent ontology, that identifies equivalent elements, but renders unrelated elements at a distance [12].

Although ontology merging can be carried out manually using ontology editor such as Protege, manual approach is difficult, labor intensive and error prone. Since Protege is our ontology building environment, we have adopted a Protege plugin called PROMPT which is Ontology Merging and Alignment Framework as our ontology merging environment. PROMPT, although is currently implemented on Protege Frames, nevertheless support most of the OWL features [13]. While merging two ontologies, PROMPT creates a list of suggested operations which are based on similarities in the class names. While merging two classes properties attached to the classes will also be merged. User can then accept or reject these suggestions [14].

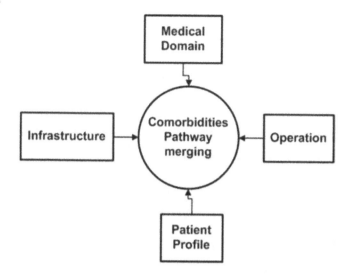

Fig. 6. Merging of pathways

4.1 Issues Related to Clinical Pathway Merging

Merging of comorbidity pathway ontologies involves reducing heterogeneity between the models. Although we are not required to resolve any syntactic heterogeneity, given both of our comorbidity pathways ontologies are encoded in OWL, we do need to address terminological and conceptual heterogeneity associated with the comorbidity ontologies (Figure 7). Terminological heterogeneity occurs when same concepts in different ontologies are referred by different names. Conceptual heterogeneity occurs when same domain is modeled differently, by either using different axioms to define the same concepts or by using entirely dissimilar concepts [13].

Since pathways are traditionally written using free text, similar medical concepts can be expressed using different terms, e.g. irregular heart rhythm can be expressed as palpitation in one pathway and arrhythmia in the pathway of its comorbidity. These semantic inconsistencies when expressed in ontology can lead to terminological heterogeneity during pathway merging. Such inconsistencies are avoided by using SNOMED CT, a standard medical vocabulary. Similarly, merging of pathways results in incidents of conceptual heterogeneity, e.g. Coronary Angiography can be modeled as either as a diagnostic procedure or a therapeutic procedure depending on the context in which it is used in the pathways.

Merging of the clinical pathways also require to avoid duplication of certain steps such as history taking, physical exam, ordering of certain lab test etc. which tends to be similar in most pathways especially for those of comorbidities. In addition, since clinical pathways are developed with specific focus in mind, their merging or alignment can be potentially dangerous. If a pathway does not explicitly state any possible drug interaction or adverse event in relation to a therapy or procedure recommended in the comorbidity pathway, the execution of merged pathway may result in wrong conclusion

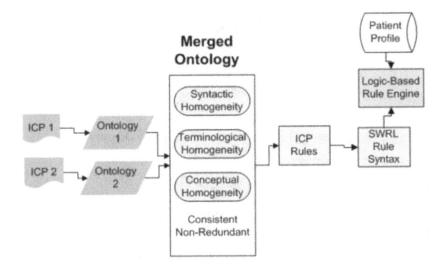

Fig. 7. Merging and execution of comorbidity pathway ontologies

or harmful recommendations such as prescription of a drug, or a procedure contraindicated in patients with a particular comorbidity.

Merging of ontologies is a highly collaborative process, particularly in medical domain, where a harmful recommendation can have disastrous consequences on the recipient of that recommendation. Avoidance of such adverse outcomes and terminological and conceptual heterogeneity require substantial collaboration with domain expert during modeling, merging and execution of the comorbidity pathways. This involves explicitly stating and discussing all the assumptions regarding the concepts and their relationships in the pathways, so that the relationships between target and source ontologies is better understood. In addition we argue that incorporating explicit descriptions of proof rules and strategies in the form of formal ontological constraints can prevent such potentially cataclysmic scenarios.

4.2 Reasoning of Merged Ontology

Once the knowledge in the comorbidities pathways is represented as a merged ontology, the subsequent stride towards realizing a Computerized Decision Support System is ascertaining a framework to represent the decision logic captured in the ontologies as IF THEN rules. A significant aspect of our approach is that computerized and merged CP can be executed, when connected with patient data, through a simple logic-based reasoning engine or a work-flow engine. In line with our Semantic Web approach we abstract SWRL (Semantic Web Rule Language) rules based on the logic captured in the merged ontology (Figure 7). These rules are of form of an inference between antecedent and consequent both containing atoms. This means that whenever the conditions specified in the antecedents hold, then the conditions specified in the consequent must also hold. These rules are written in terms of OWL classes, properties and individuals. In order to execute these SWRL rules, Protege OWL provide SWRL Engine Bridge to incorporate OWL/RDF friendly logic-based rule engines such as Racer Pro 1.9 and Jess.

Once inference has been carried out by the execution engine, the inferred knowledge is asserted back to the OWL model.

Another advantage of using SWRL for execution of the pathway ontologies is that merging of the two ontologies can also be done at rule level using SWRL. In order to map the concepts in source ontology on to the target ontology, the relevant atoms in the source ontology are used in antecedent and the corresponding atoms in the target ontology as atoms in the consequent of a SWRL rule. This approach is particularly useful in complex domains such as medical domain, where expression of mere entity definitions is not enough and more expressiveness is required.

5 Evaluation Strategy

The Decision Support System will be evaluated for modeling of the clinical domain knowledge. Ontology based on the merged pathways will be instantiated with patient records derived from the multiple hypothetical clinical scenarios depicting patients journey through the pathways while having related comorbidities. These clinical scenarios will be developed from related clinical practice guidelines, and other published and validated clinical sources and with the help of domain experts. Once the merged ontology has been reasoned about and queried, the recommendations and responses will be validated by domain experts. A short questionnaire will be prepared to measure user satisfaction of the modeling of knowledge in the system. The ethics approval will be sought before the commencement of the study.

6 Concluding Remarks and Future Work

In this paper we presented the conceptual overview of our framework for computerization and merging of clinical pathways for comorbidities to provide provide point of care decision support. Most challenging and unique aspect of this research is that it involves the dynamic integration of computerized CP of two concurrent comorbid diseases, whilst maintaining clinical pragmatics and medical correctness. Our approach is grounded in Semantic Web framework. We believe that Semantic Web have enormous prospective in computerizing and semantically linking a wide range of clinical knowledge elements in otherwise static and stand alone comorbidity pathways to implement a comprehensive and dynamic decision support system. Our Semantic Web based framework offers a platform to provide dynamic, comprehensive and more personalized monitoring and care to patients. We believe that the significance of our research is threefold. (i) Provision of point-of-care decision support for health care providers to help them make correct clinical decisions when treating patients with comorbidities, (ii) Integration of multiple CP for comorbid diseases to realize a single patient-specific care pathway guide the patients care trajectory, (iii) Knowledge translation by execution of CP into clinical practice. At the moment we are in the implementation stage of this research. We are working in the domain of Chronic Heart Failure and its comorbid conditions especially Atrial Fibrillation and Hypertension. Our work involves close coordination with cardiologists at Dalhousie University. This work is still under progress hence a complete description of implementation details and evaluation will be presented in subsequent publications.

Acknowledgements. This research is financially supported by Canadian Institute of Health Research Canada Graduate Scholarship: Doctoral Award.

References

1. Chu, S.: Reconceptualising clincal pathway system design. Collegian 8, 33–36 (2001)
2. Pearson, S., Goulart-Fisher, D., Lee, T.: Critical pathways as a strategy for improving care: Problems and potential. Annals of Internal Medicine 123, 941–948 (1995)
3. Boxwala, A.A., Peleg, M., Tu, S., Ogunyemi, O., Zeng, Q., Wang, D., Patel, V.L., Greenes, R.A., Shortliffe, E.H.: Glif3: A representation format for sharable computer-interpretable clinical practice guidelines. Journal of Biomedical Informatics 37, 147–161 (2004)
4. Tu, S., Musen, M.: Modeling data and knowledge in the eon guideline architecture. In: Medinfo 2001 (2001)
5. Tu, S.W., Campbell, J., Musen, M.A.: The structure of guideline recommendations: A synthesis. In: Proc. AMIA Symposium (2003)
6. Sutton, D., Fox, J.: The syntax and semantics of the proforma guideline modeling language. Journal of American Medical Informatics Association 10, 433–443 (2003)
7. Abidi, S.: Ontology-based modeling of breast cancer follow-up clinical practice guideline for providing clinical decision support. In: Proceedings of 20th IEEE International Symposium on Computer-Based Medical Systems. IEEE Press, Los Alamitos (2007)
8. Buchanan, B., Barstow, D., Bechtal, R., Bennett, J., Clancey, W., Kulikowski, C., Mitchell, T., Waterman, D.: Building Expert System. Addison-Wesley, Reading (1983)
9. Friedman, C., Hripcsak, G.: Natural language processing and its future in medicine: Can computers make sense out of natural language text. Academic Medicine 74, 890–895 (1999)
10. Glacer, B.: Conceptualization: On theory and theorizing using grounded theory. International Journal of Qualitative Methods 1 (2002)
11. Uschold, M., Gruninger, M.: Ontologies: Principles, methods and applications. Knowledge Engineering Review 11, 93–155 (1996)
12. Hitzler, P., Krotzsch, M., Ehrig, M., Sure, Y.: What is ontology merging?–a category-theoretical perspective using pushouts. American Association of Artificial Intelligence, http://www.aifb.uni-karlsruhe.de/WBS/phi/pub/cando05.pdf
13. Euzenat, J., Shvaiko, P.: Ontology matching. Springer, Heidelberg (2007)
14. Fridman, N., Musen, M.: Prompt: Algorithm and tool for automated ontology merging and alignment, http://smi.stanford.edu/smi-web/reports/SMI-2000-0831.pdf

Can Physicians Structure Clinical Guidelines?
Experiments with a Mark-Up-Process Methodology

Erez Shalom[1], Yuval Shahar[1], Meirav Taieb-Maimon[1], Guy Bar[2], Susana B. Martins[3], Ohad Young[1], Laszlo Vaszar[3], Yair Liel[2], Avi Yarkoni[2], Mary K. Goldstein[3], Akiva Leibowitz[2], Tal Marom[4], and Eitan Lunenfeld[2]

[1] The Medical Informatics Research Center, Ben Gurion University, Beer Sheva, Israel
{erezsh,yshahar,ohadyn,meiravta}@bgu.ac.il
http://medinfo.ise.bgu.ac.il/medlab
[2] Soroka Medical Center, Ben Gurion University, Beer Sheva, Israel
{guybar,yarkoni,liel,akival,lunenfld}@bgu.ac.il
[3] Veterans Administration Palo Alto Heath Care System, Palo Alto, CA
{Susana.Martins,Mary.Goldstein}@va.gov, laszlo@vassar.info
[4] E.Wolfson Medical Center, Holon, Israel
maromtal@013.net.il

Abstract. We have previously developed an architecture and a set of tools called the Digital electronic Guideline Library (DeGeL), which includes a web-based tool for structuring (marking-up) free-text clinical guidelines (GLs), namely, the URUZ Mark-up tool. In this study, we developed and evaluated a methodology and a tool for a mark-up-based specification and assessment of the quality of that specification, of procedural and declarative knowledge in clinical GLs. The methodology includes all necessary activities before, during and after the mark-up process, and supports specification and conversion of the GL's free-text representation through semi-structured and semi-formal representations into a machine comprehensible representation. For the evaluation of this methodology, three GLs from different medical disciplines were selected. For each GL, as an indispensable step, an ontology-specific consensus was created, determined by a group of expert physicians and knowledge engineers, based on GL source. For each GL, two mark-ups in a chosen GL ontology (Asbru) were created by a distinct clinical editor; each of the clinical editors created a semi-formal mark-up of the GL using the URUZ tool. To evaluate each mark-up, a gold standard mark-up was created by collaboration of physician and knowledge engineer, and a specialized mark-up-evaluation tool was developed, which enables assessment of completeness, as well as of syntactic and semantic correctness of the mark-up. Subjective and objective measures were defined for qualitative and quantitative evaluation of the correctness (soundness) and completeness of the marked-up knowledge, with encouraging results.

Keywords: Clinical decision support systems, Clinical guidelines, Ontologies, Knowledge acquisition, Evaluation, Mark-up, Completeness, Correctness.

1 Introduction

Clinical guidelines (GLs) have been shown to improve the quality of medical care, and are expected to assist in containment of its costs as well [1]. During the past 20

D. Riaño (Ed.): K4HelP 2008, LNAI 5626, pp. 67–80, 2009.

years, there have been several efforts to support complex GL-based care over time in automated fashion. This kind of automated support requires formal GL-modeling methods. Most of the methods use knowledge acquisition tools for eliciting the medical knowledge needed for the knowledge roles (KRs) of the GL specification *ontology* (key concepts, properties and relations) that each method assumes, in order to specify it in a formal format [2-10]. In most of those tools, however, the process of specification the textual GL into a formal language is not sufficiently smooth and transparent.

The core of the problem is that expert physicians cannot (and need not) program in GL specification languages, while programmers and knowledge engineers do not understand the clinical semantics of the GL. Therefore, we developed an architecture and set of tools, called the Digital electronic Guideline Library (DeGeL) [11], to support GL classification, semantic mark-up, context-sensitive search, browsing, run-time application, and retrospective quality assessment. DeGeL facilitates the process of specification the GL into a formal language, and enables collaboration and inherent iteration between different types of users such as: expert physicians, namely, senior, domain-expert clinicians who assist in formation of a clinical consensus that disambiguates the GL; clinical editors, namely, medically trained editors who mark-up the GL, and knowledge engineers, typically informatics experts who can create a formal GL representation.

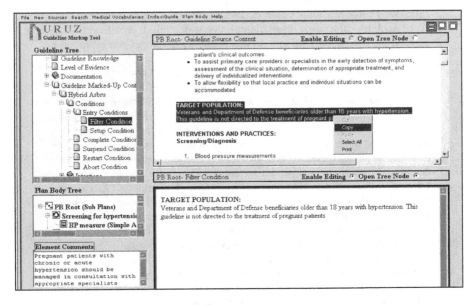

Fig. 1. The Uruz Web-based guideline (GL) mark-up tool in the DeGeL architecture. The tool's basic semi-structuring interface is uniform across all GL ontologies. The target ontology selected by the clinical editor, in this case, Asbru, is displayed in the upper left tree; the GL source is opened in the upper right frame. The clinical editor highlights a portion of the source text (including tables or Figures) and drags it for further modification into the bottom frame's Editing Window tab labeled by a semantic role chosen from the target ontology (here, the Asbru filter condition). Contents can be aggregated from different source location.

In addition, The DeGeL library supports multiple GL ontologies, in each of which, GLs can be represented in a hybrid format. One of the DeGeL framework tools is the web-based URUZ mark-up tool (Figure 1). URUZ uses the infrastructure of DeGeL's hybrid guideline representation model and thus enables expert physicians, clinical editors, and knowledge engineers on different sites to collaborate in the process of GL specification and to mark-up the GL in any representation level: semi-structured (typically performed by the clinical editor), semi-formal (typically performed by the clinical editor in collaboration with the knowledge engineer), and formal representation (typically performed by the knowledge engineer).

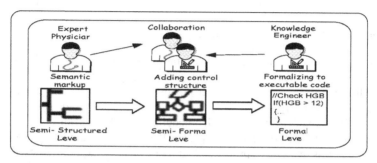

Fig. 2. The incremental specification process in the URUZ mark-up tool

This incremental process is shown in Figure 2: The expert physician indexes and mark-ups the GL (in this study, mark-up means structuring the GL text by labeling portions of text using semantic labels from chosen target GL specification language, sometimes even modifying the text into semi-formal representation), creating semi-structured representation, and in collaboration with a knowledge engineer, creating semi-formal GL representation. Then, knowledge engineers use an ontology-specific tool to add executable expressions in the formal syntax of the target ontology. Thus, each GL is represented in one or more representations levels: free-text, semi-structured text, semi-formal, and a fully structured representation. All of those GL's representation levels co-exist and are organized in the DeGeL library within a unified structure - the hybrid GL representation

2   Objectives

In this study, we aimed to answer three categories of research questions: (i) Can clinical editors actually mark-up a GL into semi-formal level, and if so, at what levels of completeness (coverage) and of soundness, or correctness (semantic and syntactic quality)? (ii) Are there significant differences in the completeness and correctness of the mark-ups between different clinical editors marking-up the same GL? (iii) Is there a significant difference between the correctness levels of the mark-up of different specific aspects of the GL?

3 Methods: The Mark-Up Process and Evaluation Methodology

The activities in the mark-up process include three main phases (Figure 3): 1) Preparations *before* the mark-up activities: choosing the specification language, learning the specification language, selecting a GL for specification, create an Ontology-Specific Consensus, acquiring training in the mark-up tool and making a Gold Standard (GS) mark-up 2) *During* the Mark-up activities: Classifying the GL according to a set of semantic indices (e.g. diagnosis, treatment), and performing the specification process using the tools and consensus, 3) After Mark-up activity: evaluation of the results of GL specification. This methodology implements the incremental specification process described in Figure 2 (particularly the structured-text and semi-formal representation levels).

3.1 Choosing the Specification Language

The first step towards specification of GL into semi-formal mark-up is to select the target GL ontology for specification (activity 1 in Figure 3). As pointed previously [13,14], different ontologies can be used for different proposes: one might select for example PRODIGY[3] for creation of common scenarios of chronic diseases, GLIF[4] or Proforma[5] for general GL modeling, or GEM[7] for documentation. In this study we used the Asbru language [12] as the underlying guideline-representation language due to its expressive procedural structures and its explicit representation: the Asbru specification language includes semantic KRs organized in KR-Classes such as Conditions (e.g., the filter condition, which represents obligatory eligibility criteria, the complete condition, which halts the guideline execution when some predefined criteria is true, and the abort condition, which aborts the guideline execution when some predefined criteria is true); control structures for the GL's Plan-Body KR-Class (e.g., sequential, concurrent, and repeating combinations of actions or sub-guidelines), GL's Intentions KR-Class (e.g. process and outcome intentions), and the Context KR-Class of the activities in the GL (e.g. actors, clinical-context).

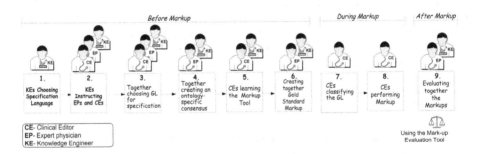

Fig. 3. The three main phases of the methodology before, during and after the mark-up, and the activities in each phase. Note the descriptions under each activity. Activity six (creation of a gold standard) can be performed in start, or in parallel with activities seven and eight (editors mark-up). Note also the participants in each activity. Note that this methodology is implements the structured-text and semi-formal representation levels which are described in Figure 2. Note step 9 is facilitated with the Evaluation mark-up tool (see section 3.8).

3.2 Learning the Specification Framework and Tools

The expert physicians and the clinical editors are instructed by the knowledge engineers about the essential concepts and aspects required for the specification process (activity 2 in Figure 3). This activity includes learning main aspects of the specification language (Asbru in our case), the hybrid model and its representation levels, the overall GL representation framework (in this case DeGeL) and its related tools. Learning the specification language will help, for example, the expert physicians to select a GL appropriate for formal representation considering its semantic aspects, and not only its clinical ones.

3.3 Selecting a Guideline for Specification

Once a decision has been made to automate a set of GLs in some clinical settings, the next step is to decide which GLs are to be formalized (activity 3 in Figure 3). A good candidate GL for specification should be one of a common disease with agreement between the majority of expert physicians on the methods of diagnosis and treatment, and with a clear, well defined clinical pathway. Thus, they might select more than one GL source when, for example, information regarding some directive is defined in more detail in another source. These sources, in addition to their own knowledge and interpretations, will serve as the basis of knowledge for creating the consensus (see Figure 4).

3.4 Creating an Ontology-Specific Consensus

In general, the textual content of the GL is not always complete or self-evident within itself: it might lack of sufficient information, suffer from ambiguousness, or require customization to local settings. Therefore, local, site-specific customizations of the GL (say, motivated by the availability of resources in the local clinical setting, by local practices, or by personal experience) must be specified explicitly to increase the probability of site-specific successful application. In addition, the same free-text GL might sometimes be interpreted differently by several local expert clinicians; in such cases, much discussion can be prevented by an explicit agreement on a common local interpretation. In their recent research Peleg et al.[15] found that in the process of making algorithms from textual GLs, teamwork is crucial for detecting errors, and that the team should include a knowledge engineer. Thus, similarly to Miller et al. [16], we too found the creation of a local clinical consensus regarding the semantics of the GL to be an indispensable, mandatory step before mark-up, and it should include both the expert physicians and the knowledge engineers (activity 4 in Figure 3).

We decomposed the crucial mark-up phase into two steps: In the first step, the local most senior expert physicians first created a clinical consensus, which was independent of any GL-specification ontology. The clinical consensus was created, for each GL, by the local senior expert physicians and a knowledge engineer. The clinical consensus was always a structured document that described in a schematic-only, but explicit, fashion the interpretation of the clinical directives of the GL, as agreed upon by the local expert physicians.

In the second step, we created what we refer to as an *Ontology-Specific Consensus* (OSC), which specifies the consensus, in terms of the chosen target ontology KRs (e.g., entry conditions). The OSC is created by senior expert physicians, who had considerable practical knowledge and experience in the relevant clinical domain, in collaboration with a knowledge engineer, who was familiar with the target specification language. Note that The OSC is a merged result of merging these two steps, and is is independent of a specification tool. An example of a free-text segment of a GL, a part of an OSC generated from that GL is in Figures 4, and 5. A detailed exposition regarding the process of creating an OSC and the evaluation of its effects can be found elsewhere [17].

Alternative Parenteral Regimens

Limited data support the use of other parenteral regimens, but the following three regimens have been investigated in at least one clinical trial, and they have broad spectrum coverage.

Ofloxacin 400 mg IV every 12 hours

OR

Levofloxacin 500 mg IV once daily
WITH or WITHOUT
Metronidazole 500 mg IV every 8 hours
OR
Ampicillin/Sulbactam 3 g IV every 6 hours
PLUS
Doxycycline 100 mg orally or IV every 12 hours.

Fig. 4. A sample of the text of the "Pelvic Inflammatory Disease" guideline's (original) source (demonstrating activity 3 in Figure 3)

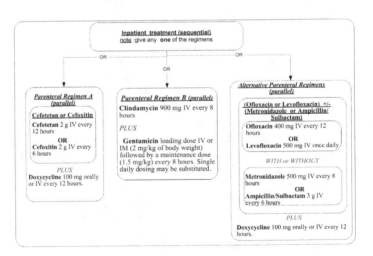

Fig. 5. A sample of the ontology specific consensus created for the "Pelvic Inflammatory Disease" guideline, demonstrating activity 4 in Figure 3. In this case, the consensus is specific to the Asbru ontology and is a small part of the consensus document that refers to the guideline shown in Figure 4.

3.5 Training the Clinical Editors in the Mark-Up Tool

The clinical editors who intend to perform the semi-formal mark-ups are instructed by the knowledge engineer in the specification tool (URUZ mark-up tool in our case) and in the OSC. A help manual or small simulation of marking-up another GL may be used to assist the clinical editor in this process of refreshing the meaning of some concepts or to confirm understanding about a task related to the specification language, the specification tool, or generally about the specification process (activity 5 in Figure 3) .

3.6 Creating the Gold Standard Mark-Up

For each of the GLs, a gold standard is created by a senior expert physician (other expert physician than the one who is participated in the mark-up activity) and a knowledge engineer together, using the OSC and GL source (activity 6 in Figure 3). The gold standard considered to be the best semi-formal mark-up and most detailed from both the clinical and semantic aspects, and therefore is used only for evaluation of the mark-ups. This step can be performed before or in parallel to the activity of performing the mark-ups.

3.7 Performing the Semi-formal Mark-Up

After the clinical editor feels enough confident, he/she can start to specify the GL using the mark-up tool (URUZ, in our case) according to the OSC and the GL sources, and create a semi-formal mark-up. In addition, the clinical editor classifies the GL according to a set of semantic indices (e.g. diagnosis, treatment) using the IndexiGuide Tool [11]. However, the knowledge engineer is not part of this session, and may help the clinical editor in case of technical problems (activities 7 and 8 in Figure 3).

3.8 Evaluation of the Mark-Ups

After the semi-formal mark-ups are completed, they are evaluated by comparing them to the gold standard according to objective completeness and correctness measures. Evaluation of the mark-up is important because it helps to qualify its quality in qualitative and quantitative measures. This stage is done by a senior expert physician and a knowledge engineer using a designated evaluation mark-up tool which enables scoring those measures for each plan, sub-plan and KR of evaluated mark-up (activity 9 in Figure 3, see next section).

3.8.1 The Evaluation Mark-Up Tool

To facilitate the evaluation, we developed a tool specialized designed to produce completeness and correctness scores: The Mark-Up Assessment Tool (MAT) [18]. The MAT is a web-based desktop application which was developed using the Dot.Net technology and enables sharing and collaboration between different sites and users. The MAT enables the expert physician, the knowledge engineer, and other guests to select and browse the desired evaluated mark-up from the DeGeL library. A typical evaluation session usually includes a physician who is an expert in the clinical domain of the evaluated marked-up GL, and a knowledge engineer who is familiar with Asbru semantics (see Figure 6).

There are two possible working modes in the MAT in each evaluation session: *View* mode and *Evaluation* mode. When one of the *evaluation managers* (usually the knowledge engineer) starts the evaluation session, he opens the MAT in the Evaluation mode, and enters the participants in the appropriate fields in the session: the names of the expert physicians, the knowledge engineers and the optional guests. In the View mode, the users can only view the mark-up. Thus, all other participants at different sites and locations in the session can open the MAT in parallel in View mode. MAT's functionality enables all online participates to see the changes made by the expert physician and by the knowledge engineer who entered in an Evaluation mode, during the evaluation session. When the evaluation manager enters the MAT, he attaches to the marked-up GL a relevant ontology-specific-consensus file. For example, for the Pelvic Inflammatory Disease (PID) mark-up, the ontology-specific-consensus of the PID GL is attached.

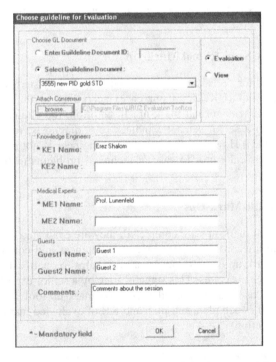

Fig. 6. The Login form of the Mark-up Assessment Tool. In the upper right frame the user can select the working mode (Evaluation or View). In the drop down list the user selects the appropriate mark-up to evaluate, and uses the "browse" button to attach an ontology specific consensus file. If the Evaluation mode is selected, the user enters the names of the knowledge engineer and the expert physician who are going to manage the evaluation session. Note the optional fields for guests.

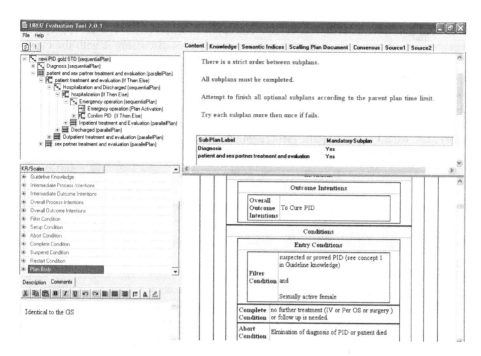

Fig. 7. The main interface of Mark-up Assessment Tool: In the upper left panel, the tree of the plans of the marked-up GL is shown. When the evaluation manager (typically the knowledge engineer) selects a plan, its procedural content (i.e., the type of plan-body, order and plans if the type is subplan, etc.) is loaded to the upper right frame, and the declarative content (i.e., all the plans' knowledge roles and their textual content) is loaded to the bottom right frame. The bottom left panel is used for giving the scores of completeness and correctness to the selected knowledge roles, according to the content quality. In addition, whenever necessary, the evaluation manager, or other participants, can use multiple tabs to view the textual source of the guideline, the ontology-specific consensus, and the guideline knowledge.

3.8.2   Assessment of the Guideline Mark-Ups Using the MAT

In the left bottom panel of the MAT main interface (Figures 7, 8), a list of checkboxes facilitates the scoring of the *completeness* and of the *correctness* measures: for each score of completeness and correctness there is a checkbox; thus, the evaluation manager, who is usually the knowledge engineer, can check the desired score in the relevant checkbox. In an evaluation session, the evaluation manager together with the expert physician opens two instances of the MAT: one for the evaluated mark-up and one for the gold-standard. Then, for each plan, starting with the root plan in the gold-standard instance, they look in the evaluated mark-up instance for a similar plan and decide on its completeness, and check the appropriate completeness checkboxes, indicating whether the plan is missing, exists or redundant, with respect to the gold-standard mark-up of that GL (see section 3.6).

In addition, the *content* of each KR of each plan (existing or redundant) is evaluated according to both syntactic and semantic measures; In this case, according to the Asbru and Clinical correctness measures. For each score [-1,0,1] (i.e, a clear error, a mild error, and a correct mark-up)of the two correctness measures there is an appropriate checkbox which the evaluation manager can check after discussing it with the expert physician . The definitions for all evaluation scores displayed as well (see Figure 8).

Thus, during the evaluation session the knowledge engineer and the expert physician are collaborating by checking the appropriate checkboxes of the completeness and correctness for each plan and KR. Additional checkboxes list each type of error, enabling them to report the error and its type (clinical semantic or Asbru syntactic error) , and what kind of specific error, again by checking the appropriate checkbox.

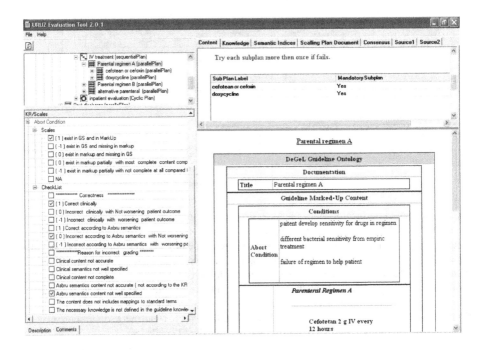

Fig. 8. Evaluation of the mark-up using MAT. Note the selected plan in the left above panel. Note the checkboxes for the completeness and correctness measures in the left bottom panel, in this case of the Abort Condition knowledge role(KR). Note also the content of the procedural content of the KR in the right above panel, and its textual content in the right bottom panel.

4   Evaluation of the Methodology

For the evaluation of this methodology, three GLs, from three distinct medical disciplines, were selected for use as the textual source for mark-ups: Pelvic inflammatory disease (PID)[19], Chronic Obstructive Pulmonary Disease (COPD) [20], and Hypothyroidism [21]. In the first stage, an OSC was created for each GL. After learning the Asbru specification language and the DeGeL framework, and receiving training in the URUZ tool, each of the clinical editors created a semi-formal mark-up using the URUZ and the OSC, and his own knowledge (for each GL two mark-ups were created by distinct clinical editors). In total six semi-formal mark-ups were created. In order to evaluate each of the mark-ups created by the clinical editors, a gold standard mark-up was created as well. Each of the mark-ups created by the clinical editors was compared to a gold standard mark-up. In addition, subjective and objective measures were defined for qualitative and quantitative evaluation of each plan, sub-plan and KR in each of the mark-ups: the subjective measures included several questionnaires to evaluate the clinical editor's attitude regarding the DeGeL framework, the specification language, the usability of the URUZ tool and their subjective benefits The objective measures were defined in two main categories: a completeness measure of the acquired knowledge, i.e., how much content from the gold standard exists (or not) in each of the semi-formal mark-ups of each clinical editor (for example, a predefined set of plans), and a correctness measure, i.e., accuracy of the acquired knowledge is according to the aspects of 1) Clinical semantics and 2) Asbru semantics. For both aspects, the score was defined according to the trichotomic scale of +1 (correct), 0 (incorrect without causing a worsening of the patient's prognosis), and -1 (incorrect and causing a worsening of the patient's prognosis). Finally, the expert physician and the knowledge engineer collaborated to perform the evaluation, using a designated graphical tool, which enables scoring the measures the mark-ups simultaneously from different sites by different users.

5 Results

A detailed analysis of the results, including the subjective measures and the measures of structuring specific KRs and a comparison of the quality of the mark-ups by distinct clinical editors for each GL is outside of the scope of this paper, which focuses mainly on the evaluation methodology. The detailed results can be found elsewhere [22,23]. However, it can be noted that the evaluation of the methodology was very encouraging when using Asbru as the specification language, DeGeL as the representation framework and URUZ as the mark-up tool. Table 1 summaries the results for all GLs: Overall, there were 196 different plans and sub-plans to be marked-up in total for the three GLs by the clinical editors: 106 plans in the case of PID GL, 59 in the COPD GL and 31 in the Hypothyroidism GL. In addition, a total number of 326 KRs were evaluated and scored in a clinical measure and an Asbru semantic measure for each plan and sub-plans in each mark-up for all GLs: 180 KRs in the case of the PID GL, 97 KRs for the COPD GL, and 49 for the Hypothyroidism GL. The completeness of the specification of all the KRs in all mark-ups was very high, with a mean of 91% ± 0.11 for all the GLs. Regarding correctness, there was quite a variability between the GLs (mean of 0.6 ± 0.7 on a scale of [-1,1]), but the quality of specification for each mark-up was always considerably higher than zero. The levels of clinical and syntactical errors were different in each GL due to the differences in the level of detail in the OSC of each GL: as the OSC was more detailed in the clinical and Asbru aspects, the correctness measures were higher. The quality of acquiring complex procedural KRs (such as cyclical plan or specification of two plans in parallel or sequential) by the clinical editors was slightly lower than acquiring declarative KRs (such as GL's filter condition).

Table 1. A summary of the completeness and correctness measures for all guideline mark-ups. Completeness is presented as a percentage; correctness is presented on a scale of [-1, 1].

		Measure - All GLs	
KR class	Number	Completeness % ± SD	Correctness [-1, 1]. ± SD
Context	47	68 ± 0.01	0.39 ± 0.92
Intentions	26	96 ± 0.04	0.89 ± 0.41
Conditions	60	88 ± 0.1	0.45 ± 0.69
Plan- Body	193	97 ± 0.02	0.66 ± 0.65
All KR Classes	326	91 ± 0.11	0.6 ± 0.7

6 Discussion

This study focused on a methodology for evaluation of a mark-up-based specification of GLs, and describes its main phases before, during and after mark-up, and the activities and participants in each phase. We presented the summarized results of a case study for this methodology, which used subjective and objective measures to compare the mark-up results to a gold standard mark-up (In practice, the gold standard can be replaced by a process of assessing the quality of the mark-up). A relative limitation is that the study used only three guidelines and six mark-up editors; however, the number of plans and subplans (196) and KRs (326) was quite high. Note that the evaluation measures can be used to analyze multiple different cross-sections: for each KR, KR-Class, all KRs in a mark-up, for each GL and for all GLs. It is also possible, for example, to classify a sample of the KRs marked-up by the clinical editors during the mark-up activity, thus supporting an ongoing quality assessment of the mark-up process [22].

7 Conclusions and Future Work

Clinical editors are capable of performing semi-formal mark-up of a GL with high completeness and correctness. However, the collaboration of an expert physician, a clinical editor, and a knowledge engineer is crucial for successful formal specification. In particular, creating an OSC is an indispensable, crucial mandatory step before mark-up. Once an OSC and a textual representation of a GL are provided, any clinical editor can in theory structure its knowledge in a semi-formal representation completely, but to specify it correctly, we should probably select a clinical editor with good computational skills or at least a solid comprehension of the GL ontology (Asbru in our case) [23]. Once we had realized the significance of the initial results of this research regarding usability and the difficulty of acquisition of procedural knowledge, we had started to develop and evaluated a new graphical interface for GL specification [24].

Acknowledgments

This research was supported in part by NIH award LM-06806. We want to thank Drs. Basso, and H. Kaizer for their efforts and contribution for this research.

References

1. Grimshaw, J.M., Russel, I.T.: Effect of clinical guidelines on medical practice: A systematic review of rigorous evaluations. Lancet 342, 1317–1322 (1993)
2. Tu, S.W., Musen, M.A.: A flexible approach to guideline modeling. In: Proc. AMIA Symp., pp. 420–424 (1999)
3. Johnson, P.D., Tu, S.W., Booth, N., Sugden, B., Purves, I.N.: Using scenarios in chronic disease management guidelines for primary care. In: Proc. AMIA Annu. Fall Symp., pp. 389–393 (2000)
4. Boxwala, A.A., Peleg, M., Tu, S., et al.: GLIF3: a representation format for sharable computer-interpretable clinical practice guidelines. J. Biomed. Inform. 37(3), 147–161 (2004)
5. Sutton, D.R., Fox, J.: The Syntax and Semantics of the PROforma guideline modelling language. J. Am. Med. Inform. Assoc. 10(5), 433–443 (2003)
6. Quaglini, S., Stefanelli, M., Lanzola, G., Caporusso, V., Panzarasa, S.: Flexible guideline-based patient careflow systems. Artif. Intell. Med. 22, 65–80 (2001)
7. Shiffman, R.N., Karras, B.T., Agrawal, A., Chen, R., Marenco, L., Nath, S.: GEM: a proposal for a more comprehensive guideline document model using XML. J. Am. Med. Inform. Assoc. 7(5), 488–498 (2000)
8. Noy, N.F., Crubezy, M., Fergerson, R.W., et al.: Protégé-2000: An Open-source Ontology-development and Knowledge-acquisition Environment. In: Proc. AMIA Symp., p. 953 (2003)
9. Votruba, P., Miksch, S., Kosara, R.: Facilitating knowledge maintenance of clinical guidelines and protocols. Medinfo. 11(Pt 1), 57–61 (2004)
10. Ruzicka, M., Svatek, V.: Mark-up based analysis of narrative guidelines with the Stepper tool. Stud.
11. Shahar, Y., Young, O., Shalom, E., Galperin, M., Mayaffit, A., Moskovitch, R., Hessing, A.: A Framework for a Distributed, Hybrid, Multiple-Ontology Clinical-Guideline Library and Automated Guideline-Support Tools. Journal of Biomedical Informatics 37(5), 325–344 (2004)
12. Shahar, Y., Miksch, S., Johnson, P.: The Asgaard project: A task-specific framework for the application and critiquing of time-oriented clinical guidelines. Artificial Intelligence in Medicine (14), 29–51 (1998)
13. Peleg, M., Tu, S., Bury, J., Ciccarese, P., Fox, J., Greenes, R.A., et al.: Comparing computer-interpretable guideline models: a case-study approach. J. Am. Med. Inform. Assoc. 10(1), 52–68 (2003)

14. De Clercq, P.A., Blom, J.A., Korsten, H.H., Hasman, A.: Approaches for creating computer-interpretable guidelines that facilitate decision support. Artif. Intell. Med. 31(1), 1–27 (2004)
15. Peleg, M., Gutnik, L., Snow, V., Patel, V.L.: Interpreting procedures from descriptive guidelines. Journal of Biomedical Informatics (2005) (in press)
16. Miller, R.A., Waitman, L.R., Chen, S., Rosenbloom, S.T.: The anatomy of decision support during inpatient care provider order entry (CPOE): Empirical observations from a decade of CPOE experience at Vanderbilt. J. Biomed. Inform. 38(6), 469–485 (2005)
17. Shalom, E., Shahar, Y., Lunenfeld, E., Taieb-Maimon, M., Young, O., Bar, G., et al.: The importance of creating an ontology-specific consensus before a markup-based specification of clinical guidelines. In: The 17th European Conference on Artificial Intelligence (ECAI2006), the workshop on AI techniques in healthcare: evidence-based guidelines and protocols, Trento, Italy, August 29. IOS Press, Amsterdam (2006)
18. Shalom, E., Shahar, Y., Taieb-Maimon, M., Lunenfeld, E.: A Quality Assessment Tool for Markup-Based Clinical Guidelines. In: AMIA Annu. Symp. Proc., Washington, DC, November 6, p. 1127 (2008)
19. Centers for Disease Control and Prevention (CDC) web site. Sexually Transmitted Diseases Treatment Guidelines (2002),
    http://www.cdc.gov/mmwr/preview/mmwrhtml/rr5106a1.htm
20. Inpatient Management of COPD: Emergency Room and Hospital Ward Management (B1), Veterans Affairs (VA) web site (2005),
    http://www.oqp.med.va.gov/cpg/COPD/copd_cpg/content/b1/annoB1.htm
21. The American Association of Clinical Endocrinologists (AACE) Web site. American Association of Clinical Endocrinologists medical guidelines for clinical practice for the evaluation and treatment of hyperthyroidism and hypothyroidism (2002),
    http://www.aace.com/pub/pdf/guidelines/hypo_hyper.pdf
22. Shalom, E.: An Evaluation of a methodology for Specification of Clinical Guidelines at Multiple Representation Levels. Master Thesis. 2006; Technical report No. 14942, Dpt. of Information Systems Engineering, Faculty of Engineering, Ben Gurion University, Beer Sheva, Israel (2006)
23. Shalom, E., Shahar, Y., Taieb-Maimon, M., Bar, G., Young, O.B., Martins, S., Vaszar, L., Liel, Y., Yarkoni, A.K., Goldstein, M., Leibowitz, A., Marom, T., Lunenfeld, E.: A Quantitative Evaluation of a Methodology for Collaborative Specification of Clinical Guidelines at Multiple Representation Levels. Journal of Biomedical Informatics 41(6), 889–903 (2008)
24. Hatsek, A., Shahar, Y., Taieb-Maimon, M., Shalom, E., Dubi-sobol, A., Bar, G., et al.: A Methodological Specification of a Guideline for Diagnosis and Management of PreEclampsia. In: The 18th European Conference on Artificial Intelligence (ECAI 2008), Workshop on Knowledge Management for HealthCare Processes, Greece, July 21. IOS Press, Amsterdam (2008)

Modeling the Form and Function of Clinical Practice Guidelines: An Ontological Model to Computerize Clinical Practice Guidelines

Syed Sibte Raza Abidi and Shapoor Shayegani

NICHE Research Group, Faculty of Computer Science, Dalhousie University, Halifax, Canada
sraza@cs.dal.ca

Abstract. Computerization of Clinical Practice Guidelines (CPG) render them to be executable at the point-of-care. In this paper, we present a knowledge modeling methodology to model the form and function of CPG in terms of a new CPG ontology that supports CPG computerization and execution. We developed a CPG ontology, in OWL using Protégé, to represent both the structural elements and the knowledge objects encapsulated in a CPG. We instantiated over 5 different CPG using our CPG ontology, whereby the instantiated CPG can be executed, with patient data, using a logic-based execution engine to provide patient-specific recommendations. We also investigated the dynamic merging of multiple CPG, at the encoding and execution levels, to handle patient co-morbidities. We evaluated the CPG ontology by examining its representational efficacy to adequately model the salient constructs of a CPG based on an existing CPG modeling formalism.

1 Introduction

Clinical Practice Guidelines (CPG) comprise a set of evidence-based recommendations to both standardize and optimize the care process, whilst ensuring patient safety and quality of care. Studies show that if CPG are integrated into the clinical workflow they reduce practice variations and costs, whilst improving the quality of care [1]. CPG are written in a free-text format, whereby they describe a set of care plans, described at different levels of abstraction, to manage a specific clinical condition. Basically, a CPG entails a set of systematically orchestrated *processes* that are applied in an episodic manner in line with the patients evolving conditions. A CPG process comprises a set of functional and temporal constraints, desired outcomes, set of actions and decision criterion. It is interesting to observe that the arrangement of processes within a CPG entails a rather intuitive and systematic structure which can be extrapolated to most CPG in general. Modeling, capturing and systematizing this 'generic' CPG structure is an interesting challenge, but it offers the potential to standardize the way we perceive the form and function of CPG, thus facilitating the computerization of CPG for execution purposes. Lately, there has been a renewed interest in computerizing CPG and incorporating them within clinical workflow to provide evidence-mediated decision support. A number of CPG representation formalisms, such as GLIF [2], EON [3], SAGE [4] and Proforma [5], have emerged with distinct approaches to computerize a CPG. However, execution of a computerized CPG is still a challenge and only a few CPG modeling frameworks offer execution of a CPG with real-life patient data.

D. Riaño (Ed.): K4HelP 2008, LNAI 5626, pp. 81–91, 2009.

Typically, most CPG conform to a generic form (i.e. structure) that is not necessarily a standard but is inherently omnipresent in most CPG and is a likely consequence of the similarities in the mental models of the CPG authors in general. For instance, most CPG include *care plans* that comprise a number of distinct *tasks* that are systematically *related* and are executed based on certain *decision criteria* and their execution follows a *temporal sequence*. Most tasks have observable *outcomes* that can be measured to determine a particular *recommendation* [4]. The presence of such implicit knowledge constructs and their systematic arrangement implies the presence of a high-level model for CPG; such a model potentially describes a systematic skeletal plan that may serve as the building blocks for a CPG. These plans are both generic and common, hence they are reusable across multiple CPG. We argue that it is both important and useful to first abstract a high-level structual model of a CPG–i.e. identify and model the key knowledge constructs, concepts, relationships, constraints and paragmatcis that are encapsulated within a CPG. In the next step, we can use the CPG model (as a template) to computerize CPG along common concepts and well-recognized relationships. It is our contention that a high-level CPG model representing the form and function of CPG can serve as the vehicle to computerize the CPG knolwedge in a standardized, re-usable and consistent manner. Potentially, there are two ways of developing a CPG model: (a) acquiring it from domain experts through interviews; or (b) inducing it by studying the knowledge artifact. We take an inductive learning approach to develop a high-level CPG model whereby we analyze a corpus of CPG to identify their constituent elements (i.e. form of CPG) and understand how these elements are used to address clinical issues (i.e. function of CPG).

In this paper, we present our methodology to abstract the underlying structural model of a CPG and represent it in terms of an ontological model–i.e. as a rich CPG ontology developed in OWL using Protégé. We present our CPG ontology that semantically models (a) the structrual, conceptual and pragmatic constructs of a CPG; (b) the domain knowledge present in the CPG; and (c) the points to merge/align multiple CPG along common steps to handle patient co-morbidities. Our CPG ontology is used to computerize CPG in a manner that they can be executed through a logic-based CPG execution engine to provide patient-specific recommendations. We establish the efficacy of our CPG ontology by (a) instantiating (i.e. computerizing) 5 different CPG; and (b) comparing its constructs with Peleg's CPG modeling formalism [6].

2 Methodology for Developing CPG Ontology

To develop our CPG ontology we take an inductive learning approach whereby we analyzed a large number of CPG (i.e. knowledge artifacts) to abstract a high-level 'semantic' model that is representative of the knowledge artifact. This abstraction is represented in terms of a CPG ontology. Our methodology comprises four steps:

2.1 CPG Classification

This step was incorporated to categorize the large body of available CPG along medical and operational dimensions. We classified the range of CPG along six dimensions as follows:

1. Severity: Acute vs. Chronic
2. Care: Primary vs. Secondary
3. Specialty Group: Medical vs. Surgical
4. Setting: Inpatient vs. Outpatient
5. Age Group
6. Orientation: Problem Oriented vs. Task Oriented

It may be noted that problem oriented CPG provide decision support for specific health-care problems, such as CPG to manage acute breathing difficulties in children, whereas task oriented CPG focus on how to perform a specific medical task, such as childhood immunization.

2.2   CPG Selection

This step involved the informed selection of multiple CPG to be used as 'exemplar' knowledge artifacts to develop the CPG ontology. The CPG classification scheme was used to objectively select a representative set of CPG to develop the CPG ontology. In total, we selected 20 different CPG that covered all the defined CPG categories. The number of CPG necessary to be analyzed was based on the premise that after analyzing a sufficient number of CPG we will reach a saturation point after which further analysis of additional CPG will not yield any new concepts. In this case, we set the initial saturation point to be 20 CPG, with the provision to select more CPG if new concepts were still being discovered after analyzing 20 CPG.

2.3   CPG Knowledge Modeling and Ontology Engineering

This step involved the analysis of the selected CPG to develop the CPG ontology. The knowledge modeling and ontology engineering was pursued based on the Model-based Incremental Knowledge Engineering (MIKE) approach that involves cyclical iterations of knowledge acquisition, model design, implementation, and evaluation [7]. In the first iteration, we conceptualized the salient CPG elements and modeled them as a preliminary CPG ontology constituting classes, attributes, and constraints. The preliminary CPG ontology was developed using the following concepts:

1. CPG metadata, such as name and description, inclusion and exclusion criteria
2. Clinical activities concerning diagnosis and treatment
3. Clinical decisions
4. Sequential organization of care activities–i.e. the modeling the order, frequency and duration of care activities
5. Clinical interventions, such as surgery and biopsy
6. Examinations
7. Medications
8. Temporal concepts

Next, we iteratively applied a middle-out approach to extend the CPG ontology. In each iteration, using the current version of the CPG ontological model, we instantiated the set of CPG selected for that iteration. In this process we extended and refined the existing version of the CPG ontology to account for new concepts, specializations and generalizations of existing concepts and alternate interpretations of existing concepts. The following tasks were performed in each iteration: (i) the set of candidate CPG were studied to extract and explicate the clinical knowledge; (ii) CPG elements were identified and analyzed, which led to either the specification of new or the refinement of existing ontology classes, attributes and constraints to model the CPG elements; (iii) Changes to the ontological model were re-evaluated to ensure semantic consistency.

By the time we reached the final iteration–i.e. working with CPG number 16 to 20, the CPG ontology had consolidated to the extent that no significant alternations/ additions were necessary. At this point, we concluded that the CPG ontology had 'saturated'–i.e. it was sufficiently expressive in its representational constructs (i.e. classes and relationships) and was deemed capable of instantiating any given CPG. At this point the ontology engineering exercise was stopped and the resultant CPG ontology was next subjected to evaluation.

2.4 CPG Ontology Evaluation

In this step, we evaluated the representational adequacy and efficacy of our CPG ontology by instantiating five new test CPG. The test CPG were selected guided by our initial CPG categorization scheme and included a diverse set of CPG. During evaluation we examined for possible ontology deletions (missing concepts), substitutions (ambiguous concepts) and insertions (superfluous concepts) that were necessary to instantiate the test CPG. In addition, we measured our CPG model using the eight dimensions of the guideline comparison proposed by [6]. Finally, we measured the ability of our CPG ontology to merge two different CPG so that they can be executed concurrently.

3 Description of CPG Ontology

Our CPG ontology represents the structural constructs and practice-oriented knowledge inherent in CPG in terms of 50 classes, 161 attributes and 589 instances. The class hierarchy is linked by the class subsumption relation, i.e. is-a relationship. Classes are denoted using UPPERCASE and attributes with *italics*. Description of all the characteristics of the CPG is not possible due to space constraints, but below we briefly describe our CPG Ontology.

The metadata (or maintenance information) for a CPG is captured by the class CLINICAL-GUIDELINE which identifies each CPG using the following attributes: *approved-by*, *author*, *authoring-date*, *comments*, *description*, *desired-outcome*, *exclusion-criteria*, *goals*, *inclusion-criteria*, *references*, etc.

In order to model the sequential execution of clinical activities suggested by a CPG, we decomposed a CPG into a set of **steps** that are followed sequentially-i.e. in order for a step to execute its preceding step must be completed. We defined a class GUIDELINE-STEP to represent the steps of a CPG (shown in figure 1). In order to model the sequence of steps in a CPG, we defined the *first-step* for CLINICAL-GUIDELINE to denote the first step in the CPG, and *next-step* attribute for each GUIDELINE-STEP instance to signify the next step that needs to be pursued. We identified three sub-classes of

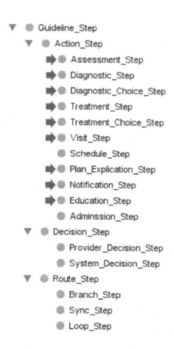

Fig. 1. CPG steps as merging points are highlighted with an arrow

the class GUIDELINE-STEP–i.e. ACTION-STEPS, DECISION-STEPS and ROUTE-STEPS, with further sub-classifications.

ACTION-STEPS represent activities performed in the CPG's workflow. We have modeled various clinical activities as sub-classes of ACTION-STEPS—i.e. ASSESSMENT-STEP, DIAGNOSTIC-STEP, VISIT-STEP, DIAGNOSTIC-CHOICE-STEP, TREATMENT-STEP, TREATMENT-CHOICE-STEP, SCHEDULE-STEP, PLAN-EXPLICATION-STEP, NOTIFICATION-STEP, EDUCATION-STEP and ADMISSION-STEP.

DECISION-STEPS represent points in the CPG where a decision needs to be made to determine the next set of activities. DECISION-STEPS are different from ACTION-STEPS because their next steps are based on the result of a decision. Their next step is modeled using the *decision-option* attribute that can hold multiple instances of *decision-option*, each instance specifying the next step that need to be taken should it be selected. We have defined two sub-classes for DECISION-STEP i.e. (a) PROVIDER-DECISION-STEP that models decisions made by the care provider, whereby its *responsible* attribute specifies the care provider who is responsible to make the decision; and (b) SYSTEM-DECISION-STEP is used when the decision-making logic is clearly specified in the CPG and in the presence of the necessary data the system can make a decision.

ROUTE-STEPS specify the flow of activities in the CPG, and have the following three sub-classes: (a) BRANCH-STEP to specify a branching point that coordinates two or more subsequent steps to be executed in parallel; (b) SYNC-STEP to synchronize (or merge) steps that have been previously branched. In order to ensure that all the steps are synchronized we have introduced an attribute *preceding-steps-to-be-completed* that ensures that all the preceding steps are completed before the control is passed to the next step; and (c) LOOP-STEP to repeat one or more guideline steps.

INTERVENTION models the set of diagnostic and treatment interventions performed during the delivery of care. There are two sub-classes: INTERVENTION-FOR-TREATMENT and INTERVENTION-FOR-DIAGNOSIS. INTERVENTION-FOR-TREATMENT represents the different types of treatment interventions, and has attributes *indication, contraindication, criteria-to-check-effects, action-if-adverse-effects*. Its sub-classes are: PRESCRIPTION, PROCEDURE-FOR-TREATMENT and RADIOTHERAPY. INTERVENTION-FOR-DIAGNOSIS represents different diagnostic interventions that are further distinguished by the following sub-classes: PROCEDURE-TO-DIAGNOSIS, DIAGNOSTIC-IMAGING GROUP-OF-DIAGNOSTIC-PROCESSES, PHYSICAL-EXAM, and LABORATORY-EXAM.

DRUG-ORDER models the type of medication(s) and their ordering information through attributes such as *drug, drug-route* and *dose-schedule*. A separate class DRUG is created to facilitate the merging of CPG (explained later) with the following attributes *allowed-roles-to-request, concept-URI* refers to the right concept in a standard medical terminology, *drug-contraindication, drug-indication, generic-name, notes-for-patient, other-names, recommended-dose* and *toxic-dose*. The DOSE-SCHEDULE captures information about the dosage of the ordered medication and the schedule for its consumption using attributes such as *dose, dose-unit* and *dose-measured*. The schedule for consuming the drug is defined by the *schedule* which holds an instance of the SCHEDULE class.

To model temporal concepts, we have defined the following two classes: DURATION that defines a time measurement value–i.e. *time-value* one week, and a measurement unit–i.e. *time-unit* with values such as Minute, Hour, Day or Week. The SCHEDULE class models different types of temporal schedules to organize activities. To specify a schedule we defined attributes such as *schedule-type, repetitions* and *duration*.

4   Using Our CPG Ontology for CPG Merging

The objective of CPG merging is to align two or more CPG to potentially handle a patient's co-morbidities which may demand the concurrent application of more than one CPG. The net outcome of CPG merging is not a new 'merged' CPG, rather the alignment of common plans/steps that exist across multiple active CPG in order to (a) realize a comprehensive decision model, encompassing multiple CPG, that targets the overall care of the patient as opposed to the treatment of just a disease, (b) optimize resources by reducing repetitive tests/actions, and (c) efficient execution of overlapping processes and interventions. Merging two (or more) CPG whilst maintaining clinical pragmatics is quite challenging because (a) recommendations that are common across multiple CPG are not necessarily administered at the same time, and (b) certain parts of the merging CPG may later result in contradictions or adverse effects. Our CPG ontology allows CPG merging at the following two levels:

Encoding level: This level approaches CPG merging during the ontology-based CPG encoding stage. The CPG ontology decomposes a CPG at the level of generic skeletal plans that can be re-used across multiple CPG. For instance, concepts such as tests and medications are usually included in multiple CPG and can therefore be defined once and then can be re-used in multiple CPG. At the encoding level, two CPG can be merged if they entail a similar plan. Figure 2 shows that CPG A and B can be merged because they both have the common step of CT-Scan. This concept is defined through the IMAGING class, which is a sub-class of INTERVENTION-FOR-DIAGNOSIS, which will have 'CT Scan' as a common instance found in both CPG A and B. In our CPG ontology, we purposely separated the INTERVENTION class from GUIDELINE-STEPS to allow medical interventions to be defined once but re-used across multiple CPG, thus serving as CPG merging points.

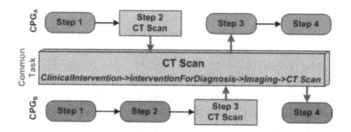

Fig. 2. CPG merging at the concept CT scan

Execution level: This level proposes the merging of common steps between multiple CPG whilst they are in execution-i.e. the execution engine looks forward for common steps and tries to merge the two concurrently running CP in order to avoid duplication of common steps. Merging CPG during execution can help eliminate repetitive steps that have long wait times, are expensive, or have potential adverse effects (i.e. radiography). Figure 1 shows the potential CPG merging points in terms of GUIDELINE-STEPS. We do not use Decision Steps as merging points because decisions are CPG-dependent and their effect is local to a CPG. Likewise, Route Steps do not qualify as merging points because they do not represent any activity, rather they facilitate the flow of activities in a CPG.

Merging CPG at the execution level needs to take into account the fact that concurrently running CPG may not necessarily have a common step to be executed at the same time. This means that the merging CPG may need to synchronize their execution, for instance one CPG may have to wait for the other CPG so that the common step can be executed, or one CPG may have to use the results of a common test done by another earlier executed CPG. We can handle three CPG merging scenarios via our CPG ontology.

Scenario 1: Both guidelines recommend a common step at the same time. Both CPG merge at the common step and then branch off to their respective paths when the common step is completed.

Scenario 2: In case the common step is not executed at the same time by two CPG, then CPG merging is still possible if the CPG in front (in terms of its execution order) can wait before executing the common step–i.e. the *ability-to-wait* constraint for the common step can be satisfied. To model this merging scenario, for each ACTION-STEP we have specified the following attributes: (a) *expected-duration* to represent the average execution time for a step; and (b) *logic-to-calculate-acceptable-wait* to specify the criteria to calculate the maximum acceptable wait time before starting the step. To estimate the length of time needed for the trailing CPG to reach the common step, the execution engine can add up *expected-duration* attributes of each step from its current state to the common step and if this time is less than the acceptable wait time for the common step, then the execution engine can withhold the execution of the leading CPG so that the execution of the common step is synchronized with the trailing CPG.

Scenario 3: Two CPG can be merged if they can re-use the results of a common step. To ensure that the result is not outdated, we have specified an attribute *acceptable-duration-of-results-if-available* that will ensure that the trailing CPG is using a valid result. If the result of a common test performed by the leading CPG is deemed outdated then the test will be repeated.

To understand CPG merging via our CPG ontology, lets assume that for a patient we need to apply two CPG: (1) Evaluation of Acute Chest Pain for Acute Coronary Syndromes and (2) Detection and Diagnosis of Hypertension. For CPG-1, a fragment of the CPG says "Consider treatment for other diagnoses if ACS is ruled out. If ACS is possible, admit patient to emergency department, perform ECG and measure cardiac markers, and decide if ACS is present". CPG-2 states "Do the following laboratory tests for patients with hypertension: Urinalysis, Blood chemistry (potassium, sodium and creatinine), Fasting glucose, Standard 12 lead ECG. Note that both CPG are recommending to perform ECG (Electrocardiogram)". An execution engine using our CPG ontology will be able to detect the common step (ECG) because both CPG will have an instance of DIAGNOSTIC-STEP in their sequence of recommendations which will indicate the need to perform an ECG. This class has an attribute *diagnostic-intervention* which holds an instance 'ECG' of the class INTERVENTION-FOR-DIAGNOSIS. Separating the intervention object from guideline steps not only allows definition of re-usable objects which leads to a smoother encoding process, but also enables the execution engine to detect that multiple CPG may require the same intervention.

5   CPG Ontology Evaluation

For CPG ontology evaluation we conducted three activities:

1. Instantiating 5 new test CPG to measure the representational efficacy of the ontology The evaluation concluded that our CPG ontology possessed the necessary representational expressiveness to instantiate the test CPG.
2. Evaluating the semantic correctness of the ontology [8]. We satisfied the three main principles relevant to ontologies–i.e. for our CPG ontology (i) each hierarchy had a single root; (ii) Non-leaf classes had at least two children; and (iii) each child was different from its parent and the siblings were different from one another.

3. Using Peleg et al. [6] framework for CPG modeling formalism as a comparator, we checked whether our CPG ontology supports the eight dimensions suggested by Peleg. This is a rather novel way of comparing a CPG against an existing modeling formalism. Below we briefly report how our CPG ontology complies with the eight structural dimensions of Peleg's formalism.

5.1   Organization of CPG Plan Components

This dimension demands the CPG modeling formalism to describe the structure of CPG plans, its components, and control flow of its processes. Our CPG ontology represents CPG as Task Network Models using distinct classes to model the core plans and components of a CPG, such as actions, decisions, and sub-plans. Our CPG ontology defines the control flow of processes using sequential, parallel or iterative activities. Earlier, we highlighted several classes to model different CPG components, most notably the class GUIDELINE-STEPS that serves as CPG plan components.

5.2   Specification of Goals and Intentions

This dimension entails the specification of the CPG goals and intentions. In our CPG ontology we are able to specify various CPG goals and intentions both as free text and formal expressions. More specifically, the CLINICAL-GUIDELINE class has two attributes for this purpose: *goals* is used to address the CPG goals as free text for user display or CPG indexing, and *desired-outcome* which expresses the desired intention of the CPG as a formal expression.

5.3   Model of Guideline Actions

This dimension concerns both the representation structure of CPG actions and how refining actions are handled if they fail to produce the intended outcomes? Our CPG ontology represents a wide range of clinical actions, such as diagnostic and treatment actions, visits to healthcare providers, communications with providers through notifications, patient education, admissions into a medical setting, and the scheduling of clinical actions. Furthermore, the CPG ontology adequately handles action refinement at various levels. For instance the INTERVENTION-FOR-TREATMENT has an attribute *criteria-to-check-effects*
which holds the criteria to check the effects of the treatment. If these criteria show that there is an adverse effect associated with the treatment, then the action specified in *action-if-adverse-effect* is to be carried out to refine the treatment. Furthermore, the class OUTCOME has an attribute *achievement-measurement-criteria* which holds the criteria to be checked if the desired outcomes of a CPG are achieved. If the criteria is met then the actions specified through *action-if-achieved* will be executed, else activities modeled by *action-if-not-achieved* will be performed.

5.4   Decision Models

This dimension concerns the presence of a definitive structure for decision constructs. In our CPG ontology, we have various instances of PROVIDER-DECISION-STEP and

SYSTEM-DECISION-STEP that serve as switch constructs because they describe mutually exclusive CPG branches that are selected based on the result of the decision making process.

5.5   Languages to Specify Decision Criteria

This dimension addresses the expression languages used to represent decision criteria, including pre- and post-conditions of CPG plan components, and the criteria that control plan execution states. For this purpose, we have created the class CONDITION with attributes *logic-text* which holds the actual logic text as an expression language, and attribute *data-elements-involved* to explicitly define the data elements used in the logic.

5.6   Data Interpretation or Abstraction

This dimension deals with the interpretation or abstractions of data elements to conceptualize CPG logic and data. An abstraction example is to use drug groups, such as ACE Inhibitor, to represent individual drugs. We have defined two classes, TREATMENT-CHOICE-STEP and DIAGNOSTIC-CHOICE-STEP to abstract general treatment or diagnostic concepts, for instance Proton Pump Inhibitor drugs are modeled through a TREATMENT-CHOICE-STEP which allows several choices for individual drugs belonging to this group.

5.7   Representation of a Medical Concept Model

This dimension deals with the ability to refer to medical terminology concepts. We support the incorporation of medical terminology concepts through attributes, such as *concept-URI* that hold the URI of the concept in the target terminology, which in turn facilitates communication between our ontology and health information systems.

5.8   Patient Information Model

This dimension deals with mechanisms to reference patient data. In our CPG ontology, patient data can also be addressed by DATA-ELEMENTS which refers to the URI of the patient data element through its *concept-URI* attribute.

In conclusion, we were able to establish that our CPG ontology is (a) sufficiently generic and expressive and generic in nature to potentially computerize any previously unseen (new) CPG with execution capabilities; (b) compliant to ontological principles; and (c) representative of key CPG constructs.

6   Concluding Remarks

In this paper we have described a knowledge modeling approach to model both the form and function of CPG in terms of a detailed CPG ontology. The CPG ontology not just captures the structural elements of a CPG but also the domain-specific knowledge held within the CPG. We recognize that our CPG ontology identifies some CPG constructs that overlap with other CPG representation formalisms, such as SAGE and EON, nevertheless we argue that our CPG ontology provides a more fine-grained classification of

CPG elements which allows for a more detailed representation and a specialized classification of concepts inherent within a CPG. Execution of a computerized CPG is made possible by the decomposition of a CPG into multiple skeletal components and interlinking them such that each action entails the next action link, thus forming a chain of actions. This approach renders the CPG execution to be modular and better tractable.

Our CPG ontology addresses the merging of concurrently active CPG. This has been made possible by defining independent and re-usable CPG objects which render CPG merging at the encoding and execution levels. We believe that CPG merging need to be pursued at the CPG knowledge level, whereby we apply high-level axioms to the domain knowledge, encoded using a CPG ontology, to achieve CPG merging that is both medically valid and clinically pragmatic.

We believe that our CPG ontology can help standardize CPG development as it can serve as a standard 'template' for knowledge explication and crystallization between health professionals engaged in a CPG authoring exercise.

Finally, we believe that this exercise of modeling the knowledge structures inherent in CPG has led to a deeper understanding of the form and function of CPG. Our CPG ontology can serve as an intermediate representation or mediator between the original CPG text and existing CPG representation formalisms.

Acknowledgement. The authors acknowledge the support of Agfa Healthcare (Canada) for funding this research.

References

1. Zielstorff, R.D.: Online practice guidelines: Issues, obstacles and future practice. Journal of American Medical Informatics Association 5, 227–236 (1998)
2. Boxwala, A.A., Peleg, M., Tu, S., Ogunyemi, O., Zeng, Q., Wang, D., Patel, V.L., Greenes, R.A., Shortliffe, E.H.: Glif3: A representation format for sharable computer-interpretable clinical practice guidelines. Journal of Biomedical Informatics 37, 147–161 (2004)
3. Tu, S., Musen, M.: Modeling data and knowledge in the eon guideline architecture. In: Medinfo 2001 (2001)
4. Tu, S.W., Campbell, J., Musen, M.A.: The structure of guideline recommendations: A synthesis. In: Proc. AMIA Symposium (2003)
5. Sutton, D., Fox, J.: The syntax and semantics of the proforma guideline modeling language. Journal of American Medical Informatics Association 10, 433–443 (2003)
6. Peleg, M., Tu, S., Bury, J., Ciccarese, P., Fox, J., Greenes, R.A., Hall, R., Johnson, P.D., Jones, N., Kuma, A.: Comparing computer-interpretable guideline models: A case-study approach. Journal of American Medical Informatics Association 10 (2003)
7. Angele, J., Fensel, D., Studer, R.: Domain and task modeling in mike. In: Proceedings of IFIP WG 8.1/13.2 Joint Working Conference (1996)
8. Bodenreider, O., Smith, B., Kumar, A., Burgun, A.: Investigating subsumption in snomed ct: An exploration into large description logic-based biomedical terminologies. Journal of Artificial Intelligence in Medicine 39, 183–195 (2007)

User-Centered Evaluation Model for Medical Digital Libraries

Patty Kostkova and Gemma Madle

City eHealth Research Centre (CeRC), City University
Northampton Square, London, UK
patty@soi.city.ac.uk

Abstract. It remains unclear whether the recent explosion of medical Internet Digital Libraries (DLs), enabled by substantial investments into eHealth by national governments and international agencies, has brought the desired improvements. As ultimately life-critical applications, medical DLs play a crucial role in delivering evidence to professionals and empowering patients. However, little attention has been given to impact evaluation with domain experts in real settings to assess whether they actually make a difference to clinical practice.

In this paper we describe a novel evaluation framework – Impact-ED – developed at CeRC to fill the gap in impact evaluation research taking into account the community, content, services and technology dimensions of DLs. We present an account of Impact-ED's application in assessing the impact of the National Resource for Infection Control in the UK (NRIC www.nric.org.uk) – a real-world medical DL used by over 40 000 professionals monthly.

Keywords: Medical Digital Library, impact evaluation, Weblogs, infection control, searches.

1 Introduction

Substantial budgets have recently been spent on eHealth programmes by national governments aiming to change the way healthcare is delivered in the 21st century [1,2]. The Internet has the potential to instantly disseminate the best available evidence, create communities of practice for professionals in widely dispersed geographical locations or for patients with rare conditions, and provide a quality-assured educational vehicle to improve the wellbeing of European and global citizens. As June Forkner-Dunn foresees: "the impact of the Internet has largely been unforeseen, and it may have a revolutionary role in retooling the trillion-dollar health care industry in the United States" [3].

However, with the growing popularity of general search engines there is an increasing need for quality-assured digital library collections which are compiled and kept up to date by experienced medical domain experts, enabling customization, personalization, and profiling of services [4]. Medical Internet DLs, ultimately life-critical applications, play a crucial role in delivering medical evidence to healthcare professionals and

D. Riaño (Ed.): K4HelP 2008, LNAI 5626, pp. 92–103, 2009.

information to the public [4]. However, is their potential fully exploited and their impact realistically evaluated?

This paper presents a novel DL impact evaluation model, the Impact-ED, and illustrates its implementation on a real-world DL, the National Resource for Infection Control (NRIC, www.nric.org.uk), hosted by the National electronic Library of Infection (NeLI, www.neli.org.uk) at the City eHealth Research Centre, City University in the UK.

Applying Impact-ED, NRIC was evaluated as a case study with infection control professionals in the UK using a combination of pre- and post-visit questionnaires, study beginning and end questionnaires, web server logs, and structured interviews.

The paper is organized as follows: Section 2 brings a background to the National Resource for Infection Control (NRIC) and outlines NRIC users' search and information needs. Section 3 discusses an overview of DL evaluation and details the Impact-ED model. In Section 4, we illustrate an application of Impact-ED on NRIC as a use case bringing detailed results while Section 5 presents conclusions.

2 National Resource for Infection Control (NRIC): Background, Web Server Traffic and Users' Information Needs

The National Resource for Infection Control (NRIC) was launched in May 2005 in response to National Audit Office [5,6] (2000/04) recommendations for a national infection control manual. The project, funded by the Department of Health in the UK and endorsed by the UK National electronic Library of Infection (www.neli.org.uk), covers a broad range of infection prevention/control and infectious diseases information and attracts over 40 000 healthcare professionals monthly.

In order to better understand users' information needs, NRIC's web server logs and traffic are evaluated on a monthly basis. In addition to basic statistics, a key area of interest is investigation of users' search terms (both internal keywords searched on the NRIC site using a built-in search engine, and external searches bringing users to NRIC from other search engines). Search keywords are a key mechanism for capturing user information needs, and feed directly into the NRIC content strategy plan.

However, do we understand users' information seeking needs and their underlying online behaviour? While there are a number of definitions of user information seeking and searching behaviour, for our purposes we shall use T. Wilson's definition of the term *information searching behavior*: all user activity on the website with the purpose of finding certain information, as opposed to "surfing" the website without a prior information need [7].

Essential indicators of the probability of knowledge discovery and overall user satisfaction and site usage are determined by

(i)   whether users use the site the way the site designers expect
(ii)   whether they understand the terminology used for site searching
(iii)   how they navigate the site to find what they are looking for

This is of particular concern in the healthcare domain, since failure to locate relevant information, or, worse, the location and use of outdated or poor quality information, can have serious consequences. Navigation can be measured by so-called "disorientation" [8] – feeling lost within the website space. Disorientation can be caused by complexity of site navigation (browsing and searching access points), unclear terminology, and poor knowledge of the domain [9].

The following section 2.1 presents a set of search results from NRIC web server logs conducted in 2008, comparing Top 20 external searches bringing users to NRIC from search engines to Top 20 internal searches performed using the NRIC internal search engine. Monthly reports providing general statistics, including geographical locations, times and dates, navigation pathways, and other important information can be found on the NRIC website[1].

2.1 NRIC Search Results

Internal search phrases are phrases input by users into the NRIC search box and processed by NRIC's built-in search engine. These are invaluable resources for understanding the information needs of users who visited NRIC (either directly or via another site or a search engine) because of a need for information about infection control.

One of the key indicators of user search behaviour is referral. External search phrases bringing people to NRIC from search engines are equally important for understanding users' needs, although their relevance relies on NRIC ranking and search engine indexing. Of all the pages viewed on NRIC, 56% were viewed by users coming from other NRIC pages, 27% from Google (google.co.uk and google.com), 0.5% from NeLI, 0.2% from traininginfection.org.uk, 0.2% from the website of the UK's Department of Health, 0.2% from www.infectioncontrol.nhs.uk, and the remainder from other websites/search engines.

External Search engine keywords in 2008. NRIC received 78 286 page views from search engines which directed users to NRIC. This is a percentage increase of 93% from 2007. The following chart, Figure 1, shows the Top 20 search phrases from online search engines. The highest ranking document called "htm 2031" provides essential NHS guidelines on infection control. The full title reads: "Health Technical Memorandum (HTM) 2031. Clean Steam for Sterilization, NHS Estates (1997)". This is one of the documents from a series of national infection control guidelines called "Health Technical Memorandum".

Internal NRIC site search keywords in 2008. In contrast, NRIC received 3 812 searches via the internet search facility on the NRIC site. This continues to confirm a trend that we have previously seen, in that browsing using the navigation menus, rather than searching, is the primary mode of finding information on NRIC. The chart on Figure 2 shows the Top 20 search phrases entered using the NRIC search facility. As shown, the Health Technical Memorandum series of resources also scores very high, while there are a number of specific keywords such as "hcai" (Healthcare Associated Infection) which did not feature on the Top 20 search engines list, probably due to the sheer volume of high-profile healthcare websites dedicated to this topical phenomenon.

---

[1] http://www.nric.org.uk/IntegratedCRD.nsf/NRIC_SiteStats?OpenForm

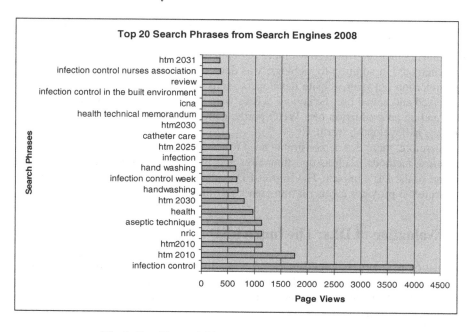

Fig. 1. Top 20 search Phrases from Search Engines 2008

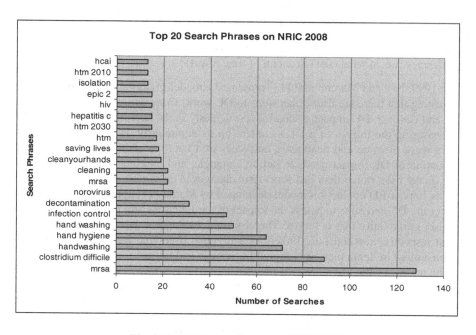

Fig. 2. Top 20 Search Phrases on NRIC 2008

While the search phrases reveal an invaluable insight into user information needs, they are only the starting point in understanding users and the actual impact of the DL on clinical practice. The need for more in-depth evaluation using multiple data collection methods, in addition to weblogs, was demonstrated by authors who illustrated a disproportion between website developers' aims and users' needs, and also between perceived and actual user behaviour. These were shown on an evaluation of NeLI DL [10] and on an evaluation of a WHO portal, called "Labresources" [11], for microbiologists in developing world.

Therefore, there is a demonstrable need to enhance the weblog results by other data gathering methods, such as questionnaires and interviews, to obtain a more rounded picture of the DL's impact. The Impact-ED evaluation model discussed in the following section 3 provides a general framework fulfilling these requirements.

3   Evaluation of DLs: The Impact-ED Model

Evaluation of Internet digital libraries is a rapidly growing research domain, of interest to the IT community as well as governments and policy makers. It has become clear to all stakeholders that sound evaluation is essential for a realistic assessment of investments into the IT technology. But what do we evaluate and what measures do we apply to assess a DL?

Chowdhury & Chowdhury define evaluation as a judgment of worth to ascertain a level of performance or value [12]. Saracevic takes this further, suggesting that performance can be broken down into two criteria [13]:

• Effectiveness, i.e. how well does a system perform that for which it was designed?
• Efficiency, i.e. at what cost (financial or time/effort)?

In 1999, Fox and Marchionini [13] presented a model of digital library dimensions, suggesting that there are four dimensions to DL work: *Community, Services, Technology,* and *Content.* DL impact evaluation, particularly in the healthcare domain, should be measuring the impact of a DL's content on its community. Is it changing a clinician's work practice and healthcare outcomes in a tangible and/or measurable way? Evaluation of DL impact needs to define, untangle, and measure the longer-term effects on the user, rather than just short-term changes in decision-making.

The Impact-ED evaluation model developed by Madle [14] applies to each dimension of the DL and aims to assess the impact of the functions and purposes of a DL on the user community that it serves. Therefore, the evaluation data, obtained from different respective methods, are linked for individual users and for the entire study to obtain a more in-depth picture of digital library use and impact.

The review of the literature [12] identified 12 healthcare DL evaluations (of which 2 were of the same DL). Only one of the studies evaluated real-time use of the DL at the point of need in the user's work, and none linked data on an individual basis from different sources. Figure 3 shows the model.

The intention of the Impact-ED (**Impact** Evaluation for **D**igital Libraries) model [14] is that a variety of methods are used to collect data, and data is linked to provide a more rounded picture of a digital library's impact.

Fig. 3. The Impact-ED model

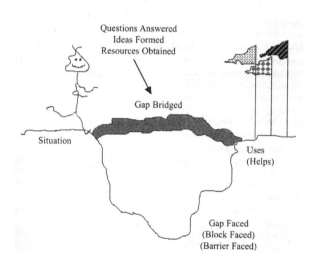

Fig. 4. Dervin's Sense-Making model (taken from [17])

The model assesses user *knowledge* gain as one of the aims of DLs, enabling users to process the data or information provided. Sharing of what is termed "explicit knowledge" (i.e. knowledge that can be written down) is considered a fundamental aim of DLs [15]. Another variable assessed is an *attitude*, defined by Azjen as "... the degree to which performance of the behaviour is positively or negatively valued ..." [16].

Are the knowledge gained and the attitudes attained by users influencing their behaviour? The Impact-ED model draws on assumptions from the Theory of Planned Behavior [16], defined by Azjen, and from the techniques of Dervin's Sense-Making model and methodology [17,18]. Dervin's approach allows exploration of *how* users meet their information needs and enables us to unravel the *"how"* of information seeking. The model is shown in Figure 4.

In order to investigate the impact of a DL from the technical, community, content, and services perspectives as defined by the Impact-ED model, appropriate data collection methods were chosen and a triangulation method developed. Reflecting the four dimensions, these are as follows:

1. *Online questionnaires.* Investigating use of the DL within the work environment
2. *Online pre- and post-visit (sense-making) questionnaires.* Investigating real-time, real-world use and how knowledge and attitudes change.
3. *Online tasks.* Investigating how users complete tasks to find information within the library and how this changes knowledge and attitudes.
4. *Weblog analysis.* Showing what users actually did within the DL.
5. *Interviews.* Complementing these other methods by providing more in-depth qualitative data that expands on issues identified in the questionnaires and weblogs.

A triangulation method, linking the pre- and post- and online questionnaires to users' actual behaviour (known from weblogs) to qualitative information revealed in follow-up interviews, provides a much more in-depth picture than previous research has allowed [18] of how a digital library may be impacting its user community and their work. The technical details of the methodology and subsequent calculation of the so-called *impact score* allowing a comparison between impacts of different DLs are too complex for the scope of this paper, and can be found in [19].

4 Case Study: NRIC Impact Assessment Using the Impact-ED

The impact evaluation of the NRIC portal was conducted using the Impact-ED model to test it in a real setting with real domain users [14]. In order to do that, the Impact-ED had to be applied on the NRIC DL to provide a set of criteria, around which questionnaires and interviews were designed to collect appropriate data. The NRIC DL was mapped onto the general Impact-ED model, to obtain insight into the impact of NRIC on the clinical practice of infection control nurses in the NHS (National Healthcare Service in the UK), as illustrated in Figure 5.

The methods used in the impact evaluation were as follows:

- *Study registration (Feb '08) and end questionnaires (May '08).* To find out how, when, why, and by whom NRIC was used, compare answers before and after the study, and provide an opportunity for users to comment on services and suggest improvements.
- *Information seeking/knowledge gain task (May '08).* To examine how well users could complete an information seeking task, i.e. find specific documents and find answers to questions using NRIC.

- *Pre- and post-questionnaire (Feb-May '08).* To discover, at the point of use and in their own words, users' reasons for using NRIC and what they know already, to compare with what they think they have learnt from using NRIC and how they will apply this to their work.
- *Web server log collection (Feb – May '08).* To collect data on how the participants actually navigate the library and see how this compares with how they report using it and the impact it has on their work.
- *Interviews (July - Aug '08).* To provide more in-depth information in users' own words about how the site has an impact and how it can be improved.

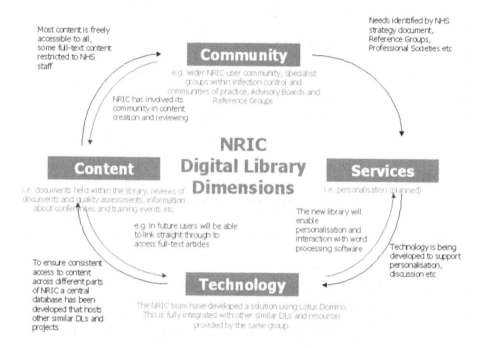

Fig. 5. Mapping the NRIC library onto the Impact-ED model

4.1 Data Analysis and Results

65 NRIC users signed up for the impact evaluation. Of these, 2 officially dropped out, 53 completed the registration questionnaire, 32 completed pre- and post-visit questionnaires of which 72 sets were matched for analysis, and 31 completed the end-of-study questionnaire. In addition, 5 users were interviewed. The study ran from February 2008 to May 2008 with interviews taking place during July and August 2008.

The majority of participants (28 of 52) listed nursing as their profession. The type of information sought at NRIC is illustrated in Figure 6.

At the start of the study most users reported the NRIC library to be either very useful (40.4%) or somewhat useful (38.5%) with only two specifically reporting that it was not useful. There was no significant change in these results at the end of the study

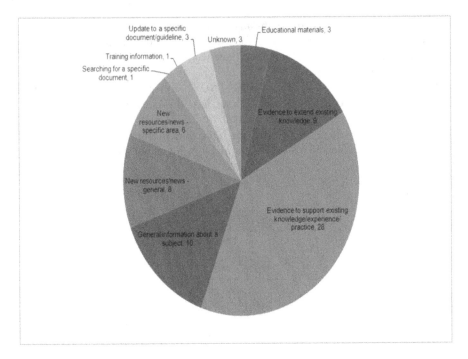

Fig. 6. Type of information sought

period. In the 72 visits for which pre- and post-visit questionnaires were collected, users found relevant information in 47 visits (65.3%).

Specific comments included:

"NRIC, um, it's a really good resource" (Interviewee A)

"I find it a very useful resource" (Interviewee B)

"Well again it just makes my job easier to do really, I think it makes me, um, it gives me the information I need to perform my role more efficiently." (Interviewee C)

Table 1 shows the basic access statistics for the 72 visits to the NRIC library that were analyzed by pre- and post-questionnaires. The average time spent (excluding the time spent completing questionnaires) was quite high: over 12 minutes, with 1/3 of users spending over 15 minutes and 1/3 under 5 minutes in the library. Users were viewing an average of 13 pages per visit and this included 3 documents. However, the majority of visits accessed either between 0 and 5 pages (41.7%) or 6 to 10 pages (34.7%).

The most popular method of navigating the website was to browse and search (27 visits), with 24 visits only browsing and 17 only searching. Browsing was more effective than searching in terms of whether or not NRIC had an impact, as shown in Table 2, perhaps due to the issues with sorting search results by date as a default, which was not implemented at the time of the study. This feature is now implemented to users' satisfaction, illustrating a direct usage of the evaluation study results on the technical improvements of the DL.

Table 1. Basic access statistics

Basic access statistics for the 72 visits analyzed	
Mean time spent per visit	00:12:22
Mean number of different pages viewed per visit	13.74
Mean number of documents viewed per visit	3.07
Median number of documents viewed per visit	2
Total number of reviews available	48
Number of reviews visited	5
% available reviews visited	10.4%

Table 2. Browsing and Searching Behaviour

Category	Confirmed/strengthened or changed knowledge	Gained knowledge	No impact
Browsed only	45.8%	50.0%	37.5%
Searched only	29.4%	23.5%	52.9%
Browsed and searched	33.3%	37.0%	48.1%

User knowledge was confirmed, strengthened, or changed in 36.1% of visits, and knowledge was gained by the user in 37.5% of visits. Most importantly, overall there was an impact on user knowledge in 52.8% of the 72 visits.

To summarize, the NRIC *Community* dimension evaluation showed that NRIC is used for policy development and for keeping up to date with news, but awareness in the community could be improved. The *Services* dimension evaluation demonstrated that NRIC was perceived as a useful resource and provided relevant information in over 65% of visits. The *Technology* dimension evaluation showed a significant amount of time spent per visit (on average over 12 minutes) and visits to 3 documents on average per session. In terms of the *Content* dimension, NRIC had an impact on user knowledge in 52.8% of visits.

As mentioned above, the next step in measuring the impact involves calculation of a single *impact score* indicating the impact of a DL's functions on an interval between 0 and 1; however, this substantial aspect of the work is beyond the scope of this paper and can be found in [19].

5 Conclusion

In recent years there has been an unprecedented explosion of medical websites and Internet DLs for patients and medical professionals. However, these vary widely in quality. Despite the massive investments in IT for healthcare, little attention has been paid to meeting the need for impact evaluation of DLs.

In this paper we illustrated the need for a user-centered evaluation of Internet medical digital libraries, taking into account the community, services, content, and technology aspects of DLs. In particular, we presented a novel Impact-ED model which brings together these four dimensions to assess knowledge, attitude, and online information-seeking behaviour to discover a DL's impact using qualitative and quantitative data collected by online and pre- and post-questionnaires, weblogs, and interviews.

The applicability of Impact-ED was illustrated in a case study undertaken on the NRIC portal in the UK in 2008 to demonstrate the impact of this infection control DL on users' knowledge, attitude, and behaviour. NRIC had a particular impact in the "Content" dimension, affecting user knowledge in 52.8% of visits. This illustrates the applicability and suitability of the Impact-ED framework to DLs in the healthcare domain.

Acknowledgements. This paper draws from a presentation given as an invited keynote talk at the Knowledge Management for Healthcare Processes Workshop, held in conjunction with ECAI 2008, presenting research undertaken by staff members at CeRC over several years. We acknowledge Sue Wiseman, Ed de Quincey, Gawesh Jawaheer, Helen Oliver, Gayo Diallo, Julius Weinberg, Anjana Roy, and Steve D'Souza for their significant contributions. Helen Oliver is also acknowledged for proof-reading this paper.

References

1. National Program for IT, http://www.npfit.nhs.uk/default.asp
2. National Knowledge Service, http://www.nks.nhs.uk/
3. Forkner-Dunn, J.: Internet-based Patient Self-care: The Next Generation of Health Care delivery. Journal of Medical Internet Research 5(2), e8 (2003)
4. Muir Gray, J.A., de Lustignan, S.: National electronic Library for Health (NeLH). BMJ 319, 1476–1479 (1999)
5. Improving patient care by reducing the risk of hospital acquired infection: A progress report (A report by the Comptroller and Auditor General). National Audit Office HC 876 Session 2003-2004, July 14 (2004),
   http://www.nao.org.uk/publications/nao_reports/03-04/ 0304876es.pdf (Summary), http://www.nao.org.uk/ publications/ nao_reports/03-04/0304876.pdf (Full Report)
6. The Management and Control of Hospital Acquired Infection in Acute NHS Trusts in England (2000), HC 230 1999/2000, http://www.nao.org.uk/pn/9900230.htm (NAO Press Notice),
   http://www.nao.org.uk/publications/nao_reports/9900230.pdf (Full Report)
   http://www.nao.org.uk/publications/nao_reports/9900230es.pdf (Executive Summary)
7. Wilson, T.D.: Human Information Behaviour. Information Science, Special Issue on Information Science Research 3(2) (2000)
8. Juvina, I., Herder, E.: The Impact of Link Suggestions on User Navigation and User Perception. In: Ardissono, L., Brna, P., Mitrović, A. (eds.) UM 2005. LNCS, vol. 3538, pp. 483–492. Springer, Heidelberg (2005)

9. Draper, S.W.: Supporting use, learning, and education. Journal of Computer Documentation 23(2), 19–24 (1999)
10. Roy, A., Kostkova, P., Weinberg, J., Catchpole, M.: Do users do what they think they do?- a comparative study of user perceived and actual information searching behaviour in the National electronic Library of Infection (2009) (submitted)
11. Madle, G., Berger, A., Cognat, S., Menna, S., Kostkova, P.: User information seeking behaviour: perceptions and reality. An evaluation of the WHO Labresources Internet portal. Informatics for Health and Social Care (in press)
12. Chowdhury, G.G., Chowdhury, S.: Introduction to Digital Libraries. Facet Publishing (2003)
13. Saracevic, T.: Digital Library Evaluation: Toward an Evolution of Concepts. Library Trends 49(2), 350–369 (2000)
14. Madle, G., Kostkova, P., Roudsari, A.: Impact-ED – A New Model of Digital Library Impact Evaluation. In: Christensen-Dalsgaard, B., Castelli, D., Ammitzbøll Jurik, B., Lippincott, J. (eds.) ECDL 2008. LNCS, vol. 5173. Springer, Heidelberg (2008)
15. Rowley, J.: The wisdom hierarchy: representations of the DIKW hierarchy. Journal of Information Science 33(2), 163–180
16. Azjen, I.: The Theory of Planned Behavior. Organ. Behav. Hum. Dec. 50, 179–211 (1991)
17. Dervin, B.: Audience as Listener and Learner, Teacher and Confidante: The Sense-Making Approach. In: Dervin, B., Foreman-Wernet, L., Lauterbach, E. (eds.) Sense-Making Methodology Reader: Selected Writings of Brenda Dervin, pp. 215–232. Hampton Press, Cresskill (2003)
18. Dervin, B.: A Theoretic Perspective and Research Approach for Generating Research Helpful to Communication Practice. In: Dervin, B., Foreman-Wernet, L., Lauterbach, E. (eds.) Sense-Making Methodology Reader: Selected Writings of Brenda Dervin, pp. 251–268. Hampton Press, Cresskill (2003)
19. Madle, G.: Impact-ED: A New Model of Digital Library Impact Evaluation. Ph.D thesis (2009)

Automatic Tailoring of an Actor Profile Ontology

Montserrat Batet[1], Aida Valls[1], Karina Gibert[2], Sergio Martínez[1],
and Ester Morales[1]

[1] Universitat Rovira i Virgili
Department of Computer Science and Mathematics
Intelligent Technologies for Advanced Knowledge Acquisition Research Group
Av. Països Catalans 26, E-43007 Tarragona, Catalonia, Spain
{montserrat.batet,aida.valls,sergio.martinez,ester.morales}@urv.cat
[2] Universitat Politècnica de Catalunya
Department of Statistics and Operations Research
Knowledge Engineering and Machine Learning group
Campus Nord, Ed.C5, c/Jordi Girona 1-3, E-08034 Barcelona, Catalonia, Spain
karina.gibert@upc.edu

Abstract. Knowledge tailoring is a powerful tool to customize a system according to the particular view and interests of the user, in such a way that usability of the system becomes significantly improved. In this paper, the tailoring of an actor profile ontology is discussed in the context of a HomeCare integrated system that is being designed and implemented in the EU K4Care project. The paper analyzes different ways of implementing the automatic tailoring of an ontology. A tool called ATAPO, which has been designed and implemented to assist the user in this automatic tailoring is presented.

Keywords: Knowledge management, Ontology, User-centered Health Care Systems, Tailoring.

1 Introduction

The K4Care is a project financed by the European Commission devoted to develop an intelligent web platform for providing e-services to health professionals, patients and citizens involved with the HomeCare (HC) of elderly patients living at home[1]. Those services aim to improve the capabilities of the new EU society to manage and respond to the needs of the increasing number of senior population requiring a personalized HC assistance. In this project, it has been defined a model of HomeCare assistance that identifies which are the common and basic home care structures (actors, services, actions, ...) shared by the main sanitary systems in Europe [1]. On one hand, this is a very valuable knowledge for defining an standard European model for HomeCare. On the other hand, establishing this model has permitted the design of an integrated system that provides facilities to the medical professionals for giving a better assistance to patients at home [2].

D. Riaño (Ed.): K4HelP 2008, LNAI 5626, pp. 104–122, 2009.
© Springer-Verlag Berlin Heidelberg 2009

This K4Care HomeCare model is formally represented by means of an ontology called Actor Profile Ontology (APO) [3]. An ontology is a structure to represent data that defines the basic terms and relationships among them and comprises the vocabulary of a topic area, as well as the rules for combining terms and relations to define extensions to the vocabulary [4]. Ontologies are widely used in medicine. These ontologies are often used to represent medical terminology, model healthcare entities with its relations, as well as to collect semantic categories of those medical concepts [5],[6].

The Actor Profile Ontology stores the knowledge about different kind of actors, including their roles, functionalities and permissions. The APO regards to the minimum elements needed to provide a basic HomeCare assistance according to the HomeCare model proposed in the K4Care project [1]. In can be enlarged with new Accessory services (oncology, rehabilitation) when required [3]. These new packages of services related to some specialized field are known as *Care Units*.

Basically, the Actor Profile Ontology indicates which are the e-health services where every type of actor can take part, the particular actions that every actor can perform as well as their data access rights. This includes the access to the Electronic Health Record of any patient or to other medical and administrative documents, often containing sensitive, private or even confidential information [7]. Moreover the APO facilitates the integration and coordination of the different actors in order to ensure a correct and efficient HomeCare assistance, since the APO clearly defines what and how care will be provided.

Since the advantage of the ontological representation over other data representation structures is to be machine readable, the contents of the APO ontology, encoded in OWL (Web Ontology Language) [8], can be used in automatic processes. In the K4Care project the APO is used to dynamically determine the behaviour of the agents under a multiagent paradigm used to implement the whole system [9]. Indeed, every particular user is represented by its own agent inside the K4Care system. When the user account is created, the code generator creates a particular agent whose behaviour is defined in the APO according to a certain actor type. In other words, the knowledge stored in the APO is used to determine the capabilities and permitted actions of every agent created into the system in such a way that the actions declared for the actor type profile in the APO are the only allowed to the corresponding user.

With this approach, the APO makes the system scalable and adaptable even to the evolutions of healthcare management policies that the APO captures or may capture in the future. The advantage of this approach is that the APO represents all the structural knowledge of HomeCare. This knowledge can be continuously updated and the system will result on an automatic seamless adaptation to the new laws, norms, or ways of doing with a null re-programming effort.

Having the behavioral knowledge integrated in a single ontology has the previously presented advantages, but also some drawbacks. Sometimes the global ontology is too big to be directly understood by a human, or part of the knowledge contained in it is only relevant for a subset of the actors or services.

Some customization process is useful, either for selecting relevant information to a particular context or even for adapting the behaviour of the system to different particular situations. This issue has been identified as playing a crucial role in the acceptance and socialization of the K4Care platform. So, it is interesting to provide some way of making a user-centered *personalization*.

From a technological point of view, it has been solved by means of a tailoring process. This tailoring consists, first of all, of eliminating from the ontology all the concepts that are not related to some particular view of the system. Thus, the tailoring is used to generate a new ontology containing a subset of the APO referred to some partial (but standard) view of the HomeCare assistance. This partial view of the general APO will be referred as *subAPO*. In a second phase, every particular user can personalize his/her subAPO according to his/her personal requirements.

As the K4Care system is based on multiagent technology and a personal agent is associated to every actor in the system[1], the behaviour of each agent is better defined by the particular subAPO corresponding to the person that the agent represents, so that it always acts according to his profile. In the architecture of the system, many different data sources are used. There is an intelligent intermediate layer, called DAL, which communicates the multiagent system with the knowledge sources, and knows, at every moment, which is the proper location of the knowledge required by the agent (APO, subAPO, Data Base, guidelines repository, etc.).

The idea of representing the actors' profile using an ontology and performing some tailoring of the global ontology to have different types of partial views can be extended to many other domains. Therefore, the content of this paper could be applicable to other areas. However, in this paper the particular case of the APO inside the K4Care system is studied.

In the K4Care system two different types of tailoring are considered:

- Tailoring by Care-Units: to extract the subAPO referred to all the knowledge of a certain Care Unit. At present the K4Care-APO contains only two Care Units: Nuclear Services and Rehabilitation Services.
- Tailoring by Actors: to extract the subAPO corresponding to a certain kind of (generic) actor (family doctor, nurse, social worker...). Once the subAPO is obtained, the tool will permit some kind of personalization of the subAPO to the particular user needs, as it will be described below.

A tool named ATAPO (Automatic Tailoring of the Actor Profile Ontology) has been designed to support the user in these tasks. In the following sections, some aspects of this tool are explained in relation to the different types of tailoring considered.

This paper analyzes different approaches to the Actor Profile tailoring and presents their advantages and drawbacks. The document is structured in the following way. In section 2, basic concepts to properly understand the paper are presented. In section 3, the care unit tailoring is explained. Section 4 presents the basic actor profile tailoring. Section 5 evaluates different mechanisms to handle the personalization of the actor subAPOs. Then, the approach taken for the K4Care project is proposed. Finally, section 6 makes a global discussion presenting some conclusions.

2   The Actor Profile Ontology in the K4Care

As said before, the K4Care Model is based on a nuclear structure (HCNS) which comprises the minimum number of common elements needed to provide a basic HomeCare service. The HCNS can be extended with an optional number of accessory services (HCAS) that can be modularly added to the nuclear structure [3]. The case of the Rehabilitation Care Unit has been studied up to now.

The distinction between the HCNS and the complementary HCASs must be interpreted as a way of introducing flexibility and adaptability in the K4Care model and also as an attempt to provide practical suggestions for standards to be used when projecting and realizing new services in largely different contexts.

All HomeCare structures (i.e. HCNS and HCAS's) consist of the same components: actors, actions, services, procedures, and documents (see Figure 1). Actors are all the sort of human figures included in the structure of HomeCare. Professional Actions and Liabilities are the actions that every actor perform to provide a service within the HomeCare structure. Services are all the utilities provided by the HomeCare structure for the care of the HomeCare patient (HCP). Procedures are the chain of events that leads one or several actors in performing actions to provide services. Information required and produced by the actors to provide services in the HomeCare structure is represented in documents.

Standard languages for codifying ontologies have been defined by the World Wide Web Consortium (W3C). The most used are OWL and RDF (Resource Description Framework) [10]. In fact, OWL is an extension of the RDF language. OWL was designed to be used in applications which require to process the contents of an ontology in an automatic way. The Actor Profile Ontology designed in the K4Care project, has been implemented in OWL, so that it can be used by any other program that supports OWL-like ontologies. Moreover, it can be visualized and modified by any OWL development tool, like the Protégé [11].

3   Tailoring by Care Unit

The K4Care APO stores the information of both HCNS and HCAS in a unique and global HomeCare ontology. However, the APO includes a concept that permits to classify some of the items according to the Care Unit to which they belong. It is called Care Unit Element (see Figure 1). An example of how some concepts inherit both from *Care Unit Element* and from the *Service* concept is given in Figure 2.

However the Care Unit classification is not complete, that is, not all the concepts inherit from this class. The concepts whose care unit reference can be

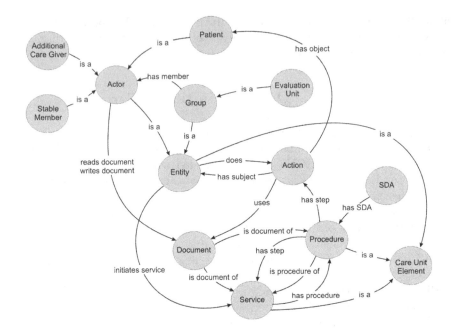

Fig. 1. The APO structure

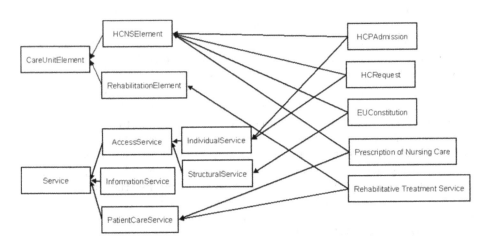

Fig. 2. Classification by Care Unit and Service

implicitly deduced do not inherit explicitly. This is done in order to avoid dupli-
cated information inside the ontology, as well as, to facilitate the enlargement of
the ontology with new HCAS, by minimizing the risk of introducing redundancies or inconsistencies.

Using the information provided in Care Unit Element, the Care Unit tailoring
consists of generating a new ontology which only contains the knowledge related

to a particular Care Unit. This process uses the classification explicitly given by the Care Unit Element class, together with the deduction of the rest of elements that are related to a particular care unit.

The sub-ontology obtained with this tailoring is a partial sight of the general APO ontology, and describes the behaviour of the HomeCare system from the particular point of view of a concrete Care Unit (it describes the kind of Home-Care assistance performed from a unit of Rehabilitation services, for example).

The subAPO can be useful for the people in charge of the corresponding Care Unit, in order to have a complete description of the unit in a structured way. This information can be useful for management purposes. For example, the subAPO indicates which kind of actors exists in the care unit, which actions perform each of them or which are the documents generated in the care unit, as well as the read/write permissions over each document for every kind of actor.

As it has been said, a tool to help in the tailoring process has been developed. This tool, called ATAPO, permits the automatic extraction of the portion of the APO referring to a certain point of view. In particular, the extraction of sub-ontologies by care unit is supported. According to the HomeCare organization model, this functionality is only available to the Physicians in Charge and Head Nurses who are responsible of the care units.

The tailoring process has been implemented in a way that it supports any modification in the content of the APO. That is, if, in the future, new elements are added or the hierarchy contains more intermediate levels, these changes will not require any modification in the ATAPO code and the sub-ontologies will be created correctly according to the new information.

3.1   Some Examples of Tailoring by Care Unit

The tailoring process by care units has been tested with the APO for HomeCare assistance, defined in the K4Care EU project [12]. Two care units have been defined and are included in the APO: HomeCare Nuclear Services (HCNS) and Rehabilitation Additional Services (HCAS-Rehab). In the ontology, the infor-mation of these two care units is merged in a way that the concepts that appear in both units are shared by both structures.

The case of the Services concept is illustrated in this section. In the K4Care model a Service is defined as a HomeCare activity that involves the work of one or more HomeCare actors in a coordinated way. They are classified into Access Services (for management), Patient Care Services, and Information Services.

The ATAPO permits to tailor the global APO to separate the information of the different care units, producing different subAPOs and storing them in separated OWL files, one for the Nuclear Services and another one for the HCAS. These partial ontologies maintain the same structure than the general APO and are also OWL-compliant.

Figure 3 shows the contents of the hierarchy of Services corresponding to the HCNS care unit, while Figure 4 regards to the Services that are involved in the HCAS-Rehabilitation unit.

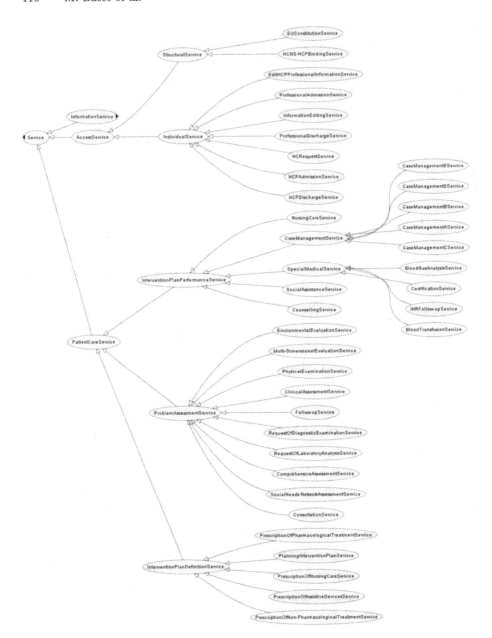

Fig. 3. Hierarchy of Services in the HCNS subAPO

Comparing Figure 3 and Figure 4, it can be seen that there are some services that are performed both in the Nuclear care unit and in the Rehabilitation care unit (e.g. a Request of Laboratory Analysis or a Planning Intervention

Fig. 4. Hierarchy of Services in the HCAS-Rehab subAPO

Plan). Some services are performed in the Nuclear Services unit but not in Rehabilitation (e.g. Social Assistance services or the EU Constitution Service, which permits to form a group of professionals (Evaluation Unit) in charge of the assessment and follow-up of a new HomeCare patient in the medical center). On the other hand, there is a particular class of services - the Rehabilitative Treatment - which is exclusive of the Rehabilitation unit and it is not provided in the Nuclear Services care unit.

With the care unit tailoring it is possible get a complete description of the performance of Rehabilitation Services care unit regarding HomeCare assistance.

4   Basic Actor Tailoring

The second type of tailoring regards to the different Actors or Groups of actors in the HomeCare model, called Entities. In this case, the tailoring process creates particular views of the APO for each type of Entity defined in the K4Care

Fig. 5. Actor hierarchy

model. The resulting subAPO describes the behaviour of this type of user in the HomeCare assistance system.

In the K4Care model, a set of HCNS actors have been defined and structured into different groups [13], distinguishing between the stable members, additional care givers and the patient. Some of the stable members may belong to an Evaluation Unit, that is a temporal group of actors in charge of the assessment and definition of an individual intervention plan of a certain patient. The K4Care model can also include other actors involved in the HCAS Rehabilitation, which were not included into the HCNS structure. Figure 5 shows the taxonomy of actors in the APO.

The user can extract a subAPO for each of the roles that he/she can play in the HomeCare assistance system. For example, Mr. Kovac, a family doctor, will be able to obtain a subAPO (with the structure shown in Figure 6) containing

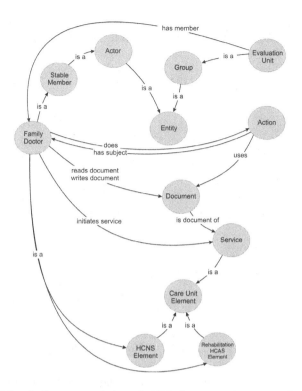

Fig. 6. Structure of the subAPO for a certain actor role

only the services, actions and documents that concerns a family doctor, and not, the ones concerning other roles (nurse, social worker).

But, if Mr Kovac is also the continuous care giver of his old and ill mother, he will also be allowed to tailor the APO under his second role of continuous care giver and then, he will get a subAPO describing the services, actions and documents concerning continuous care givers, and not those he can perform or access as family doctor.

In the same way than the Care Unit tailoring, the ATAPO tool permits the automatic extraction of the portion of the APO referring to a certain type of actor and stores it as an independent OWL ontology. This subAPO contains all the HCNS and HCAS elements that are related to the selected actor type.

The ATAPO permits to generate the subAPO of any of the roles (i.e. types of actors) defined in the K4Care HomeCare model. Furthermore, if new Actors or Groups of actors are added to the APO in the future, they will be automatically detected by the tool, and their corresponding sub-ontologies would be generated without requiring any change in the implementation. Figure 7 shows the ATAPO window where the user must select the role that he wants to tailor.

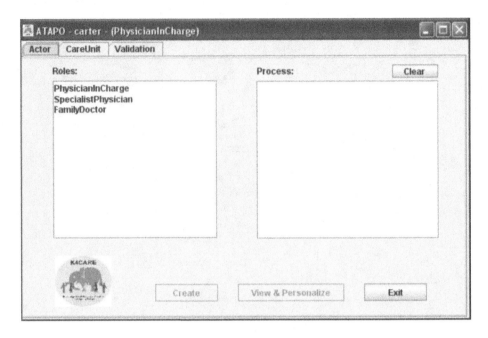

Fig. 7. ATAPO: Role selection

If other kind of modifications were done to the reference ontology (the APO), the subAPOs should be automatically rebuild, by executing again the ATAPO tool. For example, when new HCAS are added to the K4Care APO, the tailoring process must be executed again to obtain an updated subAPO of the profile.

Having the information of a particular actor profile in a separated ontology is interesting for two reasons:

1. To have a view of all the elements that are related to a particular actor profile. In fact, the actor profile is storing the way in which a certain kind of actor is participating in the HomeCare assistance [7]. This is important because, in medical domains, professional liabilities use to have legal implications. Moreover, it is needed to define an authorization mechanism for accessing to the medical information, which is considered as private and sensible data and, therefore, it is protected by privacy preserving laws. In this sense, the APO defines the information access rights for each type of actor (e.g. a radiotherapist is not allowed to read a psychological assessment scale result).
2. To have the possibility of personalizing the subAPO of a particular user. Personalization permits to increase the usability of any system because the user can customize his/her interaction with the system. However, in medical domains, it must be done in a controlled way in order to not being in conflict with the liabilities and competencies than can be taken for each type of actor. Next section is devoted to this issue.

5 Personalized Actor Tailoring

The personalization process allows the customization of a general profile for a particular user. The personalization issues to be addressed in the ATAPO have been defined by a team of medical experts together with the knowledge engineers, in order to define a process that could be consistently integrated to the K4Care system, without introducing conflicts in liabilities or service's compliance.

After a detailed evaluation of different personalization possibilities, it has been decided to provide the possibility of personalization for only two of the concepts of the APO:

- the documents that an actor can read,
- the actions that an actor can perform.

In the following sections, these two types of tailoring will be explained in more detail.

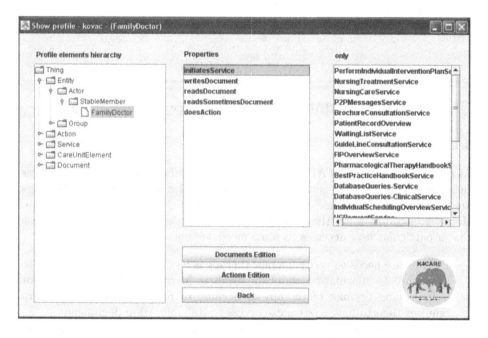

Fig. 8. ATAPO: Display of the subAPO and access to personalization tools

The ATAPO software is also being designed and implemented to support these two personalization tasks. Figure 8 shows the subAPO automatically generated by the tool for the user Mr. Kovac with the role Family Doctor. This window permits to the user to go to the documents and actions personalization wizards (Figure 9).

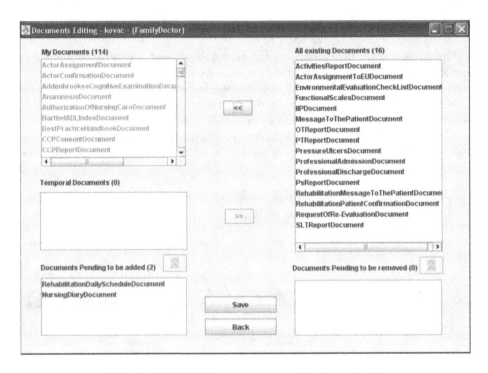

Fig. 9. ATAPO: Document personalization wizard

5.1   Documents

As said before, the global APO contains knowledge about the document's access rights for every type of actor. This means that the document's access rights are, in principle, the same for all the family doctors involved in the system. However, sometimes, in the particular context of the K4Care, it makes sense that a particular user accesses to some more documents:

- Dr Guevara needs to verify if a new therapy is having effect on his patient; to get complete information about, he needs to learn if the gait of the patient is improving because of therapy or because of rehabilitation. Dr Guevara requires then to have systematic access to Rehabilitation documents.
- Dr Kent - a Physiotherapist - realizes that Mrs Margaret remains in a stable state during the rehabilitative process; Dr Kent suspects that one of the reasons of the missing expected improvement is related to a worsening cognitive impairment and to the related drug treatment. He decides then to have systematic access to the cognitive evaluation of the patient and asks permission to read the documents reporting the pharmacological therapy.

The document tailoring consists on providing to a concrete user the possibility of consulting some documents which initially were not permitted in his profile (the complete definition of access permissions can be found in [12]). This *extension* in permissions will only provide additional Read access to those documents, but never Writing access.

The personalization of the Read access to documents will be performed in the following way:

- The ATAPO displays the list of documents initially available (with Read permission) to the user, regarding his type of actor.
- In a second region of the same window, the system displays the list of all remaining documents defined in the K4Care general APO ontology, which includes the documents used in HCNS and those used in the existent HCAS (at the moment, the Rehabilitation HCAS).
- The user can select from this second list those documents which he/she wants to access to. For example, in Figure 9 Dr. Kovac wants to have read access to the Nursing Diary and Rehabilitation Daily Schedule.
- The increase in the personal rights over the documents requires the authorization of the Physician in Charge (PC), as responsible of the safe conservation of information. Therefore, ATAPO will store the information about the documents that the user wants to access to, until the PC confirms that the request of reading access is accepted. This can proceed in two different ways:

  - Solution A: As the personal information about actors and patients is stored in a Data Base, a new table could be included in this Data Base, where the temporal additional authorizations are stored. Once the PC enters into the ATAPO and gets the list of pending authorizations, he provides them or not and the table in the Data Base is properly modified. When the user requires a certain document, the previously mentioned DAL [14] (which is in charge of retrieving information for the multiagent system) consults the particular subAPO of the querying agent, checks the table in the Data Base and only if the access is authorized provides the document.

    This is a safe way to organize the process, and the documents originally accessible to the actor are distinguishable from those temporarily permitted, but requires modification of Data Base structure and also reprogramming the DAL code accordingly.

  - Solution B: The structure of the subAPO is enlarged with respect to the APO structure, including two new relations which facilitate the management of this new kind of personalized access permissions, such as *TemporaryReadRequested* and *TemporaryReadPermitted*. The former should be activated when the user asks for additional reading rights on documents. If the the petition is authorized, a new relation (e.g. *TemporaryReadPermitted*) should be added. This solution is easy to implement since it does not modify the structure of the Data Base nor the DAL code.

- ATAPO also permits to remove the access of a document to a particular user only for those documents for which the user has previously asked additional access. This means that a particular user is not allowed to ask for not using a document which he/she is supposed to use upon the global APO description of his/her profile. For rights' removal purposes, the ATAPO only will show to the user the restricted list of additional documents previously required by the user itself. Depending on the solution used for the authorization process this removal will require an extra query to the global APO or not.

5.2   Actions

The set of possible actions to be performed by actors (professional liabilities) are strictly defined inside the K4Care model [13]. In this model, there are some actions that can be performed by more than one type of actor, as it is displayed in Table 1. In this model, agreements between actors can lead to a distribution of tasks for some period of time. In these situations, actors can personalize their subAPO accordingly. When the situation finishes, the actor must use again the ATAPO tool to change again his/her profile.

Table 1. Actions that can be performed by several types of actor

Code	Action	Actors
	Medical Activities	
M.1	perform Clinical Assessment	FD *PC SP*
M.2	perform Physical Examination	FD *PC SP*
M.3	request Diagnostic Procedures	FD *PC SP*
M.4	request Laboratory Analysis	FD *PC SP*
M.5	prescribe Pharmacological Treatment	FD *PC SP*
	Medical FD Activities	
M.FD.1	request HC	FD
M.FD.2	request of re-evaluation	FD
M.FD.3	authorize treatment proposed by SP or PC	FD
	Nursing Activities	
N.01	Specimen Collection (blood, urine, faeces)	Nu *FD PC SP HN*
N.02	I.M. Injection	Nu *FD PC SP HN*
N.03	Intravenous therapy	Nu *FD PC SP HN*

In the following, some examples of situations that can appear and make useful the previous kind of tailoring are presented. These scenarios have been provided by the medical partners of the K4Care project as some real possibilities to be considered into the system:

- Dr Kildare - a PC - is very busy in organizing the service. For these reason he decides that - for a certain period - he will perform only those actions (administration) which cannot be performed by anyone else.
- Usually doctors perform, and are expected to perform, medical actions; but they are also allowed to perform Nursing actions and, in case of need they will be allowed to do too. For example, Dr Jeckill is very busy, so he decides not to perform Nursing actions. However, Dr Zivago knows that three nurses are sick in his HomeCare center and the center have difficulties. So, he decides that he can be assigned a certain amount of specific nursing actions for his patients: during the follow-up he will check glycemia and blood pressure.

– A head nurse (HN) is expected to perform mainly organizing actions but he/she is allowed to perform Nursing actions as well. For example, HN Ms Poppins knows that for the next month the nurses will be very busy with seriously hill patients; she decides that during that period she will take care of all the catheter cares and stoma maintenances.

After an analysis of these examples, two different approaches to the personalization of actions are possible. Both approaches have some advantages and drawbacks that will be carefully detailed in the following sections.

1. The user that *lends* the action to someone else can remove them from his/her list of assigned actions.
2. The user that assumes extra actions, which are also assigned to other kind of actors, can indicate that those actions must always be assigned to him.

Tailoring by Removing Actions of the profile. This solution can be easily implemented by means of adding new relations to the subAPOs, in the same way it has been proposed in solution B for documents personalization. However, it must be considered that if all the users remove the same action from their profiles, an inconsistent situation in which an action remains with any user assigned, and therefore cannot be performed at all, can be reached. This situation can be avoided in two different ways:

A first solution (A) is to insert a control in the tailoring process in such a way that at least one person must be associated with an action. When only one person remains in charge of one action, the system blocks it and does not allow him/her to remove it from his/her actions.

Even if this solution is very easy to implement from a technical point of view and guarantees the performance of all the actions required to provide a service to the patient. However, it does not reflect the agreements in small groups of actors, as it is required in the previous examples.

A second possibility (B) is to ask the medical experts team for new information about every action, indicating which is the type (or types) of actor(s) that have to assume a certain action in their responsibilities and cannot remove this action from his/her profile. The rest of actors will have this action as optional. Actors will be only allowed to remove from their subAPO the optional actions and not the ones declared as mandatory. The optionality in the HCNS action could be the one indicated in italics in Table 1. It is only applicable to *Medical activities, Nursing Activities* and one of the *Social Activities*. In the HCAS Rehabilitation none of the actions will have any optional actor.

This solution implies an extra work from the medical partners and a modification of the APO structure, reorganizing all the defined actions in *mandatory* and *optionals* for every kind of actor. Moreover, this solution does not permit to model certain agreement between groups of users, like some of the suggested by the experts, where a couple of actors that share a same action as mandatory decide that one of them is performing it always, or the case that an actor sharing as optional a mandatory action of another actor decides to perform the action.

This is why the tailoring of the actions has been also approached on the basis of extra appropriation of actions, in front of removal of actions.

Tailoring by Appropriating Actions. In this approach, a group of people must previously agree on a task distribution (off-line) for a period of time.

Assuming that this agreement exists and the people involved will follow it, the actors must change the subAPOs accordingly. If this requirement holds, the person that wants the action must enter into the system and select the action and indicate the names of the rest of the people that will not do this action. Now we will see how the previous examples can be solved using this approach.

- Dr Kildare (PC) decides to perform only Back Office Activities. He must then delegate the Nursing Activities and Medical Activities to his colleagues. He must find a family doctor (FD) or specialist physician (SP) that uses the ATAPO to take those Medical Activities and remove them from Dr Kildare profile, and the same with Nursing Activities.
- Dr Jeckill (FD) decides not to perform Nursing actions. A nurse must use ATAPO to take that actions and remove them from Dr. Jeckill subAPO.
- Dr Zivago (FD) decides that he can be assigned a certain amount of specific nursing actions for his patients: during the follow-up he will check glycemia. In this case, Dr Zivago uses ATAPO to select action N.15 and to indicate which particular Nurse (Nu), PC, SP and HN are relieved from this task, which will be the ones in his medical team.
- Ms Poppins (HN) decides that during a period she will take care of all the catheter cares and stoma maintenances. She will select those actions in the ATAPO and indicate which Nu, FD, PC and SP (the ones of her team) will not do those tasks any more.

After a certain period of time, the actors that were relieved from some actions should use ATAPO to incorporate them again into his/her profile (the tool will control that only actions that are defined in the general K4Care APO are allowed).

In this approach it is not possible that an actor does actions that are out of this scope. It is totally avoided to have actions without any actor assigned. For these reasons, it is not required to have any authorization process in this type of personalization. But it can be used for safety, and it can also be useful to control that every actor reassigns properly their own actions when the agreement period finishes.

However, this solution will only work properly if the agreements between users are achieved off the platform (by conversations etc) and the implied persons correctly introduce in the ATAPO the proper modifications. Since the ontologies do not include information about concrete person, there are some issues that will not be supported in the ATAPO tool:

- The system will not make any control over the periods of times in which the agreements are valid. That is, no timing information will be managed automatically. The system should be properly updated by the users as soon as the agreements change.

- Different agreements in different groups of actors (e.g. Evaluation Unit (EU)) will not be considered.
- Agreements between pairs or groups of actors with respect to particular patients will not be allowed.

This approach is adaptable and supports situations as the one suggested by the experts. It is easy to maintain and to implement. So, it also could be a possible way of implementing the personalization.

6 Discussion and Conclusions

In this paper different approaches to the tailoring and personalization of actor profiles represented with ontologies have been presented. For the case of Home-Care assistance a tool to support this task has been implemented, called ATAPO. In this domain, two types of personalization are proposed: read access to extra documents, modification of the set of actions that can be assigned to a user. The paper proposes different approaches to sort them out in a semi-automatic way. The advantages and drawbacks of each proposal are analysed. Having into account that in medical assistance there are quite strict norms and laws to follow, it is mandatory to include authorization processes that prevents the users to personalize their profile without notifying it to the responsible of the care unit.

After this study, it is worth to note that ontologies are not the most suitable way for representing dynamic knowledge as personalization. Ontologies are a much better support to represent static general knowledge about a domain. This means that none of these solutions is optimal, although they provide some flexibility.

Although the use of subAPOs by the personal agents is a good way to implement a dynamic behaviour model for the multiagent system, it seems not appropriate to store particular personalizations regarding a temporary change in the user permissions or liabilities. Since the personalization is dynamical and associated to a particular person and do not contain structural information of the system, probably the best solution is to store this temporal information in a Data Base, together with the personal information about the user. This is compatible with the use of subAPOS if the subAPO is regenerated automatically every time that the user logs into the system, considering the possible intermediate updates of the global APO and the personal customization indicated in the Data Base for the user. However, this requires the modification of the data storage structure of the K4Care system, which is no currently possible, because the Data Base has been designed to basically store the Electronic Health Care Record.

Including the personalized tailoring in the K4Care system requires some cautions regarding the inclusion of new HCAS to the main APO. Since the addition of new HCAS in the general APO may enlarge the actions of the current actors of the system (for example, in the situation where a hospital opens a new oncology service involving part of the current staff of the hospital), the process to propagate these new activities to the personalized subAPOs of the corresponding actors is quite complex, and implies to solve a problem of merging ontologies, which is reported as a future line of work.

Acknowledgments. This work has been funded by the K4Care project (IST-2004-026968) and by the Student Research Grant of the University Rovira i Virgili. The authors also acknowledge the help of the doctor Fabio Campana.

References

1. Campana, F., Moreno, A., Riaño, D., Varga, L.: K4care: Knowledge-based home-care e-services for an ageing europe. In: Annicchiarico, R., Cortés, U., Urdiales, C. (eds.) Agent Technology and e-Health. Whitestein Series in Software Agent Technologies and Autonomic Computing, Switzerland, pp. 95–115. Springer, Heidelberg (2008)
2. Isern, D., Moreno, A., Pedone, G., Varga, L.: An intelligent platform to provide home care services. In: Riaño, D. (ed.) K4CARE 2007. LNCS, vol. 4924, pp. 149–160. Springer, Heidelberg (2008)
3. Gibert, K., Valls, A., Casals, J.: Enlarging a medical actor profile ontology with new care units. In: Riaño, D. (ed.) K4CARE 2007. LNCS, vol. 4924, pp. 101–116. Springer, Heidelberg (2008)
4. Neches, R., Fikes, R., Finin, T., Gruber, T., Senator, T., Swartout, W.: Enabling technology for knowledge sharing. AI magazine 12(3), 36–56 (1991)
5. Pisanelli, D.M.: Ontologies in Medicine. Studies in health tehnology and informatics, vol. 102. IOS Press, Amsterdam (2004)
6. Open Biomedical Ontologies,
   http://www.bioontology.org/tools/portal/bioportal.html
7. Gibert, K., Valls, A., Lhotska, L., Aubrecht, P.: Privacy preserving and use of medical information in a multiagent system. In: Advances in Artificial Intelligence for Privacy Protection and Security. World Scientific, Singapore (2009) (in press)
8. Dean, M., Schreiber, G.: OWL Web Ontology Language Reference. W3C Working Draft (2008), http://www.w3.org/TR/owl-ref/
9. Hajnal, A., Moreno, A., Pedone, G., Riaño, D., Varga, L.: Formalizing and leveraging domain knowledge in the K4CARE home care platform. In: Semantic knowledge management: an ontology-based framework, pp. 279–302. Idea Group publisher, IRM Press, CyberTech Publishing and Idea (2008)
10. Fensel, D.: The semantic web and its languajes. IEEE Intelligent Systems 15(6), 67–73 (2000)
11. Noy, N.F., Fergerson, R.W., Musen, M.A.: The knowledge model of protégé-2000: Combining interoperability and flexibility. In: Dieng, R., Corby, O. (eds.) EKAW 2000. LNCS, vol. 1937, pp. 17–32. Springer, Heidelberg (2000)
12. Gibert, K., Valls, A., Casals, J.: D04.2 - sample apos. Internal deliverable for the K4CARE project (2007)
13. Campana, F., Annicchiarico, R., Riaño, D.: D01 - the k4care model. Internal deliverable for the K4CARE project (2007)
14. Batet, M., Gibert, K., Valls, A.: The data abstraction layer as knowledge provider for a medical multi-agent system. In: Riaño, D. (ed.) K4CARE 2007. LNCS, vol. 4924, pp. 87–100. Springer, Heidelberg (2008)

A Methodological Specification of a Guideline for Diagnosis and Management of PreEclampsia

Avner Hatsek[1], Yuval Shahar[1], Meirav Taieb-Maimon[1],
Erez Shalom[1], Adit Dubi-Sobol[2], Guy Bar[2],
Arie Koyfman[2], and Eitan Lunenfeld[2]

[1] The Medical Informatics Research Center, Ben Gurion University,
Beer Sheva, Israel
{hatsek,yshahar,meiravta,erezsh}@bgu.ac.il
[2] Soroka Medical Center, Ben Gurion University, Beer Sheva, Israel
{sobolad,guybar,ariek,lunenfld}@bgu.ac.il

Abstract. There is a broad agreement regarding the necessity of building usable and functional tools for specification of machine-interpretable clinical guidelines, to provide guideline-based medical care. There is much less of a consensus of how to go about it and whether the whole process is feasible in a real clinical domain. In this study, we have applied a new architecture, the Gesher graphical tool, to the specification of an important Obstetric guideline (diagnosis and management of PreEclampsia / Eclampsia toxemia). We have assessed the feasibility (functionality and usability) of (1) representing a clinical consensus customized for a particular medical center and (2) structuring the full content of the guideline. In addition, we have assessed in a preliminary fashion, the potential of using a less experienced clinician as a markup editor, by asking both a senior Obstetrics and Gynecology clinician, and a general intern, to represent the same guideline using the Gesher system. The results demonstrated the functionality and usability of the Gesher system, at least for these two editors; the intern's performance was at least as good as that of the senior physician with respect to the specific task of structuring the guideline according to the Hybrid Asbru ontology, using our tools.

Keywords: Clinical Guidelines, Clinical Protocols, Care Plans, Knowledge Representation, Knowledge-Based Systems, Medical Decision Support Systems, Obstetrics and Gynecology, Medical Informatics, Evaluation.

1 Introduction

Over the past 20 years, there were multiple efforts to provide decision support for medical care by formalizing clinical guidelines (GLs) into machine interpretable formats [1, 2]. GLs represented in a machine-comprehensible format, can be applied by a computerized agent as a tool to support physician decisions at the point of care, or as a tool for retrospective quality assessment and research. Several guideline-specification ontologies such as GLIF [3], Proforma [4], and Asbru [5] were developed to represent guidelines in a formal and machine interpretable format. In this paper we present the

D. Riaño (Ed.): K4HelP 2008, LNAI 5626, pp. 123–133, 2009.
© Springer-Verlag Berlin Heidelberg 2009

Gesher system, wich is a software tool designed to support guideline specification into multiple representation levels. We also present a preliminary evaluation of the Gesher system that was conducted in the specification of an Obstetric GL for diagnosis and management of the PreEclampsia / Eclampsia disease.

2 Graphical Specification of Clinical Guidelines

The Gesher system is a graphical client application designed to support the process of incremental guideline specification at multiple representation levels. Gesher is part of the DeGeL [6, 7] framework which is a comprehensive framework that includes a web-based GL library and a suit of tools for GL specification, retrieval, visualization and application. DeGeL uses a hybrid representation, namely a representation methodology that includes several intermediate and final formats that are stored within the knowledge base. The current representation formats include (1) the original full text, (2) a structured representation (marked-up text), (3) a semi-formal representation which includes control structures such as sequential or parallel sub-plans order, and (4) a fully formal, machine-comprehensible format. The intermediate representation levels have additional benefits; the semi-structured level is crucial for context-sensitive search of GLs; the semi-formal level supports application of the GL by a clinician at the point of care, without access to an electronic medical record.

Gesher is designed to support the gradual knowledge acquisition process performed by a collaboration of an Expert Physician (EP) familiar with the domain-specific clinical knowledge, a Clinical Editor (CE) familiar with general medical knowledge and with the semantics of the target GL ontology, and a Knowledge Engineer (KE). This collaboration, which includes specification of the guideline ontology specific consensus, is critical for achieving high quality specification [8]. Gesher supports the gradual process through all the steps of creating formal representation of the guideline and for further maintenance and modification of the guidelines. The methodology we used for the specification process consists of the following major steps:

- Formation of a clinical consensus – Expert physicians customizing the guideline and deciding on the recommendation to be adapted from the source guidelines; the customizations may include changes to fit the local clinical setting.
- Creation of a structured consensus – Expert physicians and knowledge engineers collaborate to specify the knowledge of the guideline according the selected specification ontology.
- Markup: The clinical editor is creating a complete structured representation of the guideline, according the structured consensus.
- Specification of the semi-formal and formal representation by the knowledge engineer. Currently the semi-formal and formal specification uses the Asbru ontology [5]. Asbru is a GL representation language that enables GL specification using a hierarchical (plan, sub-plan) representation and an expressive temporal specification language for Actions (e.g. periodic plans), Data Abstractions (e.g. clinical temporal patterns), Conditions (i.e. conditions to enter or abort or complete a plan) and Intentions (e.g. process, outcome).

2.1   Specification of the Structured Consensus

Gesher supports the composition of the GL from plans and sub-plans that represent the GL's procedural control flow. This procedural control flow represents the clinical algorithm that is recommended by the GL. The EPs and the KEs use the graphical components of Gesher to create the hierarchical structure of the GL. These components include graphical flow-chart diagrams and hierarchical tree-structure to explore the plans hierarchy (Figure 1). The hierarchical structure of the GL is stored in De-GeL along with elements of procedural and declarative knowledge that are stored for each sub-plan. When creating the structured consensus the user inserts new sub-plans to the GL by dragging plans from several existing semantic types (such as clinical procedure, observations or drugs). When the user chooses to declare a certain plan as composite, a new diagram is created where the procedural type will be defined (e.g. sequential or parallel application of the sub-plans).

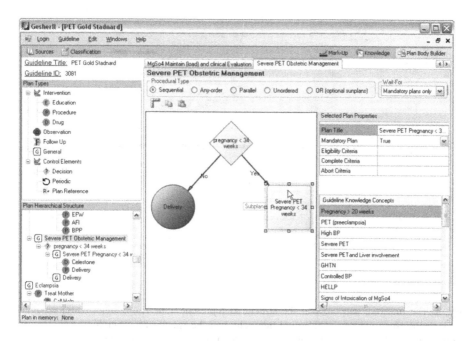

Fig 1. The Hierarchical Plan Builder in Gesher used by the expert physicians to create the structured consensus of the GL

2.2   Specification of the Structured Representation

After the structured consensus is created, the CEs use Gesher to create complete structured representation of the GL (a mark-up, structuring of free text process). The structured representation is an intermediate representation and not yet a formal one, but is crucial for achieving the formal representation in gradual process. Each plan and sub-plan is constructed from elements according to the knowledge roles of the selected target ontology (e.g. abort-condition or clinical-settings). In order to create a structured representation the CE needs to refine the plans, created in the structured consensus, and link those plans to portions of text from the original text-based GL, where the evidence-based recommendations are obtained. To edit the structured representation content (Figure 2), Gesher provides an HTML editor and other graphical components to support the user when performing the markup. The user navigates through the complex hierarchical structures of the GL, the structure of the specification ontology, and then performs the markup by dragging portions of labeled content from one or more source guidelines into the selected knowledge roles frames. The back pointers to the source text are saved in the GL library to allow the display of the areas in the source GLs from where the text of each knowledge-role is originated.

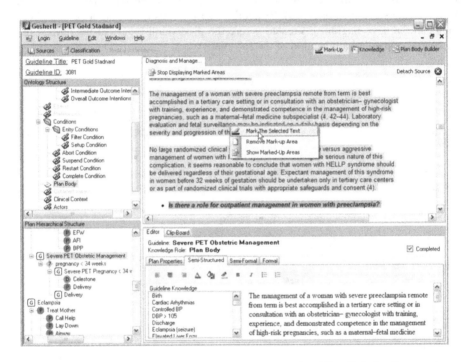

Fig 2. The interface used to edit the structured representation. The clinical editor navigates through the structures of the GL plan hierarchy and the specification ontology and then performs markup of text to create content for each of the plan's knowledge roles.

2.3 Specification of the Semi-formal and Formal Representation

The Semi-Formal representation is a more intermediate representation of the GL and defers from the structured representation mainly by a more formal description of expressions within the GL. We are currently experimenting intuitive graphic controls that we developed in order to allow the EPs and CEs to be involved in the more formal specification. Involving physicians in the specification of the semi-formal representation should help the process of creating the formal representation by extracting more of the required knowledge for creating a fully machine comprehensible format.

The semi-formal representation includes the logical and temporal relations between the elements which are composing the expressions in the GL. The semi-formal representation can be used by the runtime application engines for applying the plans using the system's user (e.g. physician or nurse) as a mediator to the patient data [9].

The Formal representation of a GL is obtained in order to execute the GL, by a computer agent, to perform tasks such as decision support or retrospective quality assessment; therefore, it should include a formal definition for each aspect of the GL. While dealing with the control flow and logical expression in the intermediate representation levels, in the Formal level, it is required to create a complete definition for executing the expressions within the GL, for a given patient. These expressions include requirements for patient data, and in many GL scenarios, will consist of temporal aspects.

To accomplish the task of representing at formal level the declarative knowledge with temporal aspects, Gesher uses the interface of the temporal abstraction knowledge acquisition tool, which is one of the tools used in the IDAN[10] architecture.

3 The Preeclampsia / Eclampsia Toxemia Guideline

To evaluate the Gesher system and the GL specification methodology, we have conducted a study that included the specification of the GL for diagnosis and management of PreEclampsia and Eclampsia Toxemia (PET) [11]. PET is a pregnancy disease characterized by high blood pressure and appearance of protein in the urine and is highly dangerous both to the fetus and to the mother. PET occurs in approximately 12–22% of pregnancies, and is directly responsible for 17.6% of maternal deaths in the United States. We have created a structured representation of the PET GL, which includes the required customizations for the application of this GL in a real clinical setting, in this case, the OB/GYN ward of the Soroka Medical Center.

4 Objectives

The objectives of our research were to evaluate the feasibility, namely the functionality and usability, of the Gesher system in supporting the tasks of structuring the GL consensus and performing the mark-up process. We were also interested in assessing, in a very preliminary fashion, the possibility of succeeding in these tasks when performed by a CE who is a senior expert physician, and is an expert in the GL's domain, compared when performed by a CE who is an intern and has only general medical knowledge. We were also interested in assessing the time and human efforts required for completing the specification process, and in discovering the common errors and mistakes, committed during the specification process.

5 Methods

In following section we describe the methodological specification of the PET GL. Several participants took place in the process: (1) the expert physician (EP) and the knowledge engineer (KE) who created together the ontology specific consensus that was structured in Gesher. They also created the gold standard specification in Gesher that was used in the evaluation. (2) the senior OB/GYN expert physician (expert) and a general intern (intern) who participated as the Clinical Editors (CEs) that used Gesher to create the structured consensus and the structured representation of the GL. (3) a group of senior expert physicians that took part in creating the clinical consensus based on to the recommendations of the ACOG guideline for diagnosis and management of PreEclampsia and Eclampsia.

5.1 Creating a Clinical Consensus

After choosing the GL for the specification we have conducted two meetings with senior expert physicians from the Soroka Medical Center. In these meetings, they created a clinical consensus, which included the recommendations from the original source guideline and modifications and additions required for the customization and implementation of the GL in this specific medical institution.

5.2 Creating Ontology Specific Consensus

Following the clinical consensus achieved in the first step, the EP, in collaboration with the KE, have created the Ontology Specific Consensus (OSC), which is a detailed document describing both the procedural aspects of the clinical algorithm and the declarative definitions of the medical concepts within the GL (e.g. the criteria for diagnosing severe PET). The OSC was created as a printed flow-chart format.

5.3 Creating Structured Consensus in Gesher

The CEs used Gesher to structure the clinical flow chart specified in the OSC. The structured consensus includes all the sub-plans composing the GL and specifying the semantic type of each sub-plan such as drug or observation. The procedural type of each composite plan was determined in this step (e.g. sequential or parallel) and each plan was defined as mandatory or optional, with respect to being completed in order to allow the process to continue to the next sub-plan. For each sub-plan several properties were defined as structured text (e.g. "abort condition"). The plan properties included optional annotations describing the following procedural aspects: the eligibility criteria to enter the plan, the conditions for completing the plan successfully or for aborting the execution of the plan, and in the case of periodic (reaping) plans, a specification of aspects such as frequency and number of iterations. Another task performed in this step was the creation of the list of all medical concepts defined in the OSC. This list is called the GL Knowledge and was later further refined by the physicians.

5.4 Creating a Structured Representation in Gesher

The CEs were provided with the GS structured consensus and than used Gesher to create the complete structured representation of the GL according to the Knowledge-Roles (KRs) of the Asbru specification ontology. Each of the sub-plans in an Asbru based GL is composed from KRs such as "abort-condition" and "clinical-settings". The physicians used Gesher to perform markup of text from the source GL, to create these KRs for each of the sub-plans. The reference to the source was saved in the GL library. Although the structured consensus includes informal specification of all KRs, in this study the markup phase included only the following KRs: (1) Filter-Condition (2) Setup-Condition (3) Complete-condition (4) Abort-Condition (5) Actors (6) Clinical-Context.

6 Evaluation

The structured consensus and the structured representation created by the intern and by the expert were compared to a Gold Standard (GS) structured consensus and a GS structured representation. The GS were created by the EP together with the KE and were assumed to be the most detailed and correct specifications, clinically and semantically. The comparison to the GS was also performed by the EP together with the KE; using software tool was built especially for this purpose. The interface of this evaluation tool is presented in figures 3 and 4, using it in the evaluation enabled the users to navigate through the complex structure of the guidelines which include more than 115 sub-plans, and to compare and grade each part of the structured consensus and the structured representation.

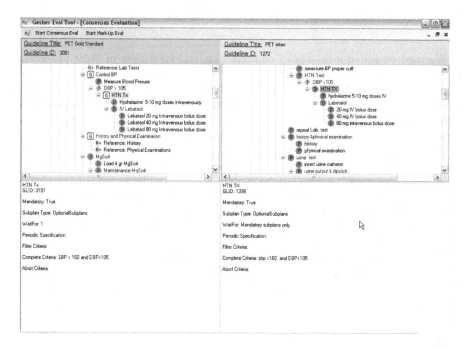

Fig 3. The interface of the evaluation tool, used to navigate through the complex structure of the guidelines and to compare each sub-plan to the gold standard specification. The gold standard is displayed on the left side of the screen and the specification to evaluate is displayed on the right side.

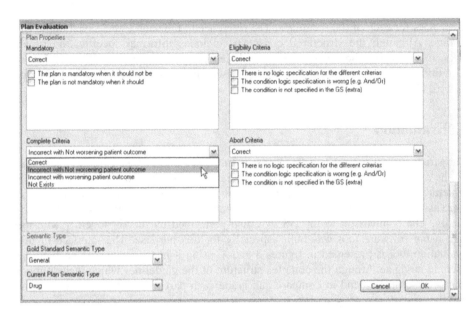

Fig 4. The interface of the evaluation tool, used to grade the procedural elements comprising each sub-plan

We used objective measures of completeness and soundness in order to measure the level of success of the physicians to complete the specification tasks. The completeness of the specification was defined as the number of knowledge elements (such as sub-plans or plan properties) that exist in the structured consensus and the structured representation created by the physicians compared to the number of knowledge elements that exist in the GS. In order to measure the soundness of the specification products, each of the knowledge elements was assigned a discrete grade describing its correctness; if the content of the element was correct and similar to the GS, the element was assigned as "Correct", if the content was not similar to the GS, the element was assigned one of the following grades; "Not correct worsening the patient outcome" or "Not correct not worsening the patient outcome". The overall level of correctness of a guideline or a group of KRs is the proportion of knowledge elements assigned with the grade "Correct".

To evaluate the usability of the Gesher, we used the SUS [12] questionnaire that was presented to the users after performing each of the tasks. To assess the time efforts, we measured the time required for the expert and the intern to complete the specification process.

7   Results

7.1   Results for Creating a Structured Consensus in Gesher

We started by assessing the feasibility of structuring the existing OSC document, by the two CEs using Gesher. Both the expert and the intern have achieved high level of completeness in this task. The proportions of plans that exists both in the GS structured consensus and in the structured consensus, were 96% and 97% for expert and the intern respectively. The proportions of completeness of the plan properties of each of the existing sub-plans (e.g. periodic specification and complete condition) were 99.6% and 100% for the EP and the CE respectively.

The soundness of the structured consensus created by the physicians was measured by the proportion of the knowledge elements that were scored as "Correct" in the evaluation. The expert and the intern have achieved very high level of correctness, 97.78% and 99.22% respectively. Table 1 summarizes these results, the knowledge elements are partitioned into two classes; the procedural class includes the elements of procedural type (e.g. parallel or sequential), mandatory specification and periodic specification, the condition class includes the elements of the eligibility criteria, complete and abort conditions.

Table 1. The proportion of correct knowledge elements in the specification of the structured consensus in Gesher by the two CEs

	Procedural Type Knowledge Elements	Condition Type Knowledge Elements
Expert	97.56%	98.64%
Intern	98.59%	100.00%

7.2 Results for Performing the Markup in Gesher

Both CEs have performed the markup based on the GS structured consensus that was provided to them. Table 2 summarizes the results for the completeness and correctness of the structured representation. The KRs of the Asbru specification ontology, that were used in the evaluation, partitioned into three classes; Condition KRs (e.g. complete, abort), Context KRs ("actor" and "clinical-settings") and Knowledge definition KRs.

Table 2. The proportion of the complete and correct KRs in the specification of the structured representation of the PET GL

KRs	Expert		Intern	
	Complete	Correct	Complete	Correct
Conditions	467/468 (99.79%)	462/467 (98.93%)	468/468 (100%)	466/468 (99.57%)
Context	148/234 (63.25%)	150/150 (100%)	221/234 (94.44%)	220/220 (100%)
Knowledge	11/18 (61.11%)	9/11 (81.18%)	17/18 (94.44%)	15/17 (88.24%)

The completeness of the structured representation that was created was defined as the proportion of the KRs with complete content from all KRs of all sub-plans. Both the EP and the CE achieved a very high level of completeness in structuring the conditional KRs, 99.79% and 100% respectively. The "actors" and "clinical-context" KRs were specified at a lower level of completeness, 63.25% and 94.44%, by the expert and the intern respectively.

The soundness of the structured representation was measured by the proportion of KRs that were judged as being similar to the GS and to have the correct (clinically and semantically) content. Both CEs achieved high level of soundness for structuring all classes of KRs.

7.3 Referencing the Source Guideline

Another interesting measure we examined is the completeness of the references from the KRs of each sub-plan to the text of the source GL. Each of the knowledge elements can include back pointers to the text of the source GL where the recommendations originated. It is interesting to note in both markups the low proportion (about 50%) of elements that actually exist in the source GL. To explain the result in Table 3, the expert structured 278 KRs, from which 134 have back pointers to the source according to the GS. He completed full references to 107 KRs, partial reference to 4 KRs and missing reference to 23 KRs.

Table 3. The completeness of references to the source GL

Expert	Full	107/134 (79.85%)
	Partial	4/134 (2.99%)
	Missing	23/134 (17.16%)
	Exist in source	134/278 (48.2%)
Intern	Full	184/193 (95.34%)
	Partial	0/193 (0%)
	Missing	9/193 (4.66%)
	Exist in source	193/355 (54.37%)

7.4 Results for the Usability of Gesher

The following results for assessing the usability of the interface of Gesher were achieved by using the standard system usability survey (SUS) that was presented to the CEs after performing each task. Mean score of 85% for the interface for structuring the consensus, and a mean score of 77.5% for the interface for performing the markup. The Gesher tool was thus considered quite usable by both CEs.

7.5 Time Efforts

The overall time efforts for completing both parts of the specification process were measured for both CEs. The expert has worked for 32 hours (that where separated into 16 shorter episodes) and the intern has worked for 46 hours.

8 Discussion

The Gesher system was found reasonably functional and usable in supporting the specification of the structured consensus and the structured representation. It is encouraging fact that the general intern had successfully completed the tasks in levels of completeness and correctness that are not inferior to the senior physician. To better understand how only 50% of the knowledge elements that were marked-up exist in the source GL text, it is important to note that in the current methodology that we used for the GL specification, much of the knowledge was created by the expert physician in the phase of creating the consensus, and could not be pinned down to a specific text in the source GL.

References

1. Peleg, M., Tu, S.W., et al.: Comparing Computer-Interpretable Guideline Models: A Case-Study Approach. Journal of American Medical Informatics Association 10(1), 52–68 (2002)
2. De Clercq, P., Blom, J., et al.: Approaches for Creating Computer-Interpretable Guidelines that Facilitate Decision Support. Artificial Intelligence in Medicine 31(1), 1–27 (2004)
3. Boxwala, A.A., Peleg, M., Tu, S., Ogunyemi, O., Zeng, Q.T., Wang, D., et al.: GLIF3: a representation format for sharable computer-interpretable clinical practice guidelines. J. Biomed. Inform. 37(3), 147–161 (2004)
4. Fox, J., Johns, N., Rahmanzadeh, A.: Disseminating medical Knowledge: the PROforma approach. Artificial Intelligence in Medicine 14(1), 157–181 (1998)
5. Shahar, Y., Miksch, S., et al.: Asgaard project: A Task-Specific Framework for the Application and Critiquing of Time-Oriented Clinical Guidelines. Artificial Intelligence In Medicine 14(1-2), 29–51 (1998)
6. Shahar, Y., Young, O., et al.: A Framework for a Distributed, Hybrid, Multiple-Ontology Clinical-Guideline Library and Automated Guideline-Support Tools. Journal of Biomedical Informatics 37(5), 325–344 (2004)
7. Hatsek, A., Shahar, Y., et al.: DeGeL: A Clinical-Guidelines Library and Automated Guideline-Support Tools. In: Ten Teije, A., Miksch, S., Lucas, P. (eds.) Computer-based Medical Guidelines and Protocols: A Primer and Current Trends. Studies in Health Technology and Informatics, vol. 139, IOS Press, Amsterdam (2008)
8. Shalom, E., Shahar, Y., et al.: A Quantitative Assessment of a Methodology for Collaborative Specification and Evaluation of Clinical Guidelines. J. Biomed. Inform. 41(6), 889–903 (2008)
9. Young, O., Shahar, Y., et al.: Runtime Application of Hybrid-Asbru Clinical Guidelines. J. Biomed. Inform. 40(5), 507–526 (2007)
10. Boaz, D., Shahar, Y.: A Framework for Distributed Mediation of Temporal-Abstraction Queries to Clinical Databases. Artificial Intelligence in Medicine 34(1), 3–24 (2005)
11. American College of Obstetricians and Gynecologists: Diagnosis and Management of Preeclampsia and Eclampsia. Washington (DC): American College of Obstetricians and Gynecologists (ACOG), January 2002, 9 p. (ACOG practice bulletin; no. 33) (2002)
12. Brooke, J.: SUS - A quick and dirty usability scale (1996),
    http://www.usability.serco.com/trump/documents/Suschapt.doc

Home Care Personalisation with Individual Intervention Plans

David Isern[1], Antonio Moreno[1], Gianfranco Pedone[2], David Sánchez[1],
and László Z. Varga[2]

[1] University Rovira i Virgili
Department of Computer Science and Mathematics
Intelligent Technologies for Advanced Knowledge Acquisition Research Group
Av. Països Catalans, 26. 43007 Tarragona, Catalonia (Spain)
{david.isern,antonio.moreno,david.sanchez}@urv.cat
[2] Hungarian Academy of Sciences
Computer and Automation Research Institute
Kende u. 13-17. 1111 Budapest, Hungary
{gianfranco.pedone,laszlo.varga}@sztaki.hu

Abstract. The adoption of general intervention plans (guidelines) according to the particular circumstances of both the patient and the doctor's diagnostic is a very challenging task. Although there are some available geriatric guidelines, the usual Home Care patients suffer from a set of co-morbid conditions and diseases that hampers the direct application of standard protocols. The representation, creation and execution of individual intervention plans in a Home Care unit are complex tasks to be accomplished by different kinds of actors. This paper presents an architecture for dealing with the whole Home Care scenario but paying special attention to two issues. The first is the creation of Individual Intervention Plans by merging several clinical guidelines. The second is the accurate description of medical knowledge in order to know which data are contained in every step of the pathway and which actor is able to perform it.

Keywords: Home Care, ICT, intelligent agents, ontologies, intervention plans.

1 Introduction

In *e*-Health it is increasingly necessary to develop computerised applications to support medical practitioners [1]. The care of chronic and disabled patients involves lifelong treatments under continuous expert supervision. Such patients are beginning to saturate national health services and increase health-related costs all around Europe. This challenge can be faced through ubiquitous assistance (*Home Care -HC- model*). The K4Care European project proposes a new model of HC to support the provision of care services to a patient that requires assistance at home. The typical HC Patient (HCP) is an elderly patient, with co-morbid conditions and diseases, cognitive and/or physical impairment,

D. Riaño (Ed.): K4HelP 2008, LNAI 5626, pp. 134–151, 2009.

functional loss from multiple disabilities, and impaired self-dependency [2]. The healthcare of the HCP is particularly complex because of the great amount of resources required to guarantee a quality long-term assistance.

In the geriatric field there are some clinical guidelines available [3,4], which suggest the best treatment for some conditions; however, nearly 50% of adults older than 65 years have 3 or more chronic medical conditions [5], and the clinical guidelines of each individual pathology cannot be directly applied. It is widely accepted that the use of standardised guidelines brings several benefits both to patients and medical practitioners, including, among others, the assurance of a high level of care, the use of clinical knowledge about the patient at the appropriate point of care, and the security that the treatment is based on results scientifically proven in controlled randomised clinical trials, thereby reducing errors [6,7]. However, it is also widely known that there are many problems that hamper the daily use of standard guidelines by practitioners, such as a lack of awareness with the guideline's existence, lack of agreement, lack of physician self-efficacy, lack of outcome expectancy, or the inherent difficulty to change habits in daily behaviour [8].

In this scenario, the computerised agent-based platform developed in the K4Care project addresses not only the definition of fully personalised *Individual Intervention Plans* (IIPs), that take into account both the medical and social characteristics of the patient and the medical knowledge available in standardised geriatric clinical guidelines (known as *Formal Intervention Plans*, FIPs [6]), but also its execution. The intelligent and autonomous agents of the multi-agent system coordinate their activities in order to ensure an automatic and efficient execution of all the steps included in the treatment of a particular patient [2,9].

This paper focuses on the construction and execution of IIPs. First the *K4Care platform architecture* is described. After that we comment the SDA* formalism, which is the one used to represent both FIPs and IIPs within the system [10]. The paper then turns to the description of how IIPs are designed by medical practitioners and how they are executed by the platform. Finally, several conclusions and potential lines of further work are presented.

2   Home Care Platform Architecture

In the case of elderly people with chronic illnesses, it is widely accepted that a home treatment increases their life quality and reduces costs. While the healthcare information system of a medical centre may be rather centralised, Home Care assistance naturally needs a distributed service-oriented architecture, as many kinds of actors from different institutions are involved. Thus, the K4Care platform needs to manage information from distributed sources. The K4Care platform architecture has three main modules: the *Knowledge Layer*, the *Data Abstraction Layer*, and the *K4Care agent-based platform* (see Figure 1) [2].

The *Knowledge Layer* includes all the data sources required by the platform. It contains an Electronic Health Record (EHR) subsystem that stores patient records with personal and medical data. The *Actor Profile Ontology*

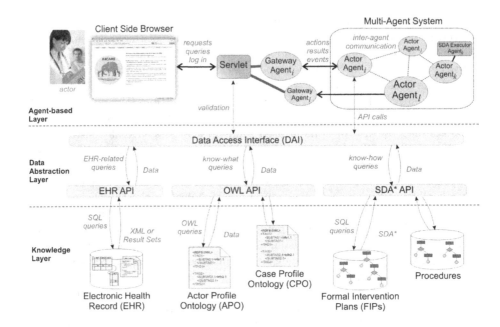

Fig. 1. K4Care Platform Architecture

(APO) contains information about all the types of actors allowed in the system (*e.g.*, actions that each type of actor can perform). The *Case Profile Ontology* (CPO) includes medical information on syndromes, symptoms and diseases (*e.g.*, symptoms associated to each disease). Both ontologies are represented in OWL [11]. The platform provides two databases with procedural information codified in SDA* ([10], see sec. 3.1). There is a repository of the procedures that implement administrative services, and a store of FIPs, which are general descriptions, usually defined by international care organisations, of the way in which a particular pathology should be dealt with.

The *Data Abstraction Layer* (DAL) provides Java-based methods that allow the K4Care platform entities to retrieve the data and knowledge they need to perform their tasks. It offers a wide set of high-level queries that provide transparency between the data (knowledge) and its use (platform) [12].

The *K4Care platform* is a web-based application. Any actor interacts with the system through a Web browser and is represented in the system by an *Actor Agent* that knows all details about his roles, permissions, pending results, pending actions, and that manages all queries and requests coming from the user or other agents. In order to exchange information between the agents and the actors there is an intermediate bridge constituted by a servlet and a *Gateway Agent* (GA). The servlet is connected with the browser's user session. When an actor logs into the system, the servlet creates a GA whose mission is to keep a one-to-one connection with the corresponding permanent agent. In this manner, the *Actor Agent*'s execution is independent of the user's location. Ide-ally, *Actor Agents* should be executing in a secure environment and distributed through several nodes in order to provide load balancing in conditions of high concurrence.

3 Home Care Personalisation

3.1 Representation of Procedural Intervention Plans

All the procedural knowledge related to (Formal and Individual) Intervention Plans is represented in K4Care using the SDA* language, which is basically a flowchart-based representation of clinical guidelines [10]. The most basic com-ponents of SDA* structures are the domain variables, which can be of three types:

a) *State variables*, that represent terms that are useful to determine the condi-tion of the patient at a certain stage (*e.g.*, a patient with motor impairment of the arm).
b) *Decision variables*, which are required by medical experts to choose among alternative medical, surgical, clinical or management actions within a treatment (*e.g.*, a treatment may follow different paths depending on the spasticity of the patient).
c) *Action variables*, which represent the medical, surgical, clinical or manage-ment actions that may appear within a treatment (*e.g.*, make trunk control exercises).

These items are linked with directed edges, which can be labelled with tem-poral constraints in order to represent deadlines or periodic actions. Figure 2 shows an SDA* structure that represents the treatment of post-stroke rehabili-tation. Each FIP may have several entry points, which indicate different points in which the treatment may start, depending on the initial state of the patient. States are represented as circles, decision as rhombus and actions as rectangles. The patient is initially in the start entry point, and an evaluation of his level of motor impairment is made, using standardised scales. After calculating the Tinetti index [13], and if the patient cannot walk, a set of muscular exercises should be prescribed by practitioners. If the patient does not present spasticity and has improved his mobility, at the final state of the treatment he can walk. In other branches of the FIP physicians might use other tools such as medication, surgery or the use of a wheel chair.

3.2 Definition of an Evaluation Unit for Each Patient

In the K4Care platform, FIPs may be related to a syndrome (*e.g.*, cognitive impairment), a symptom (*e.g.*, abdominal pain) or a disease (*e.g.*, dementia). We assume that the patient (HCP) may suffer from several of these conditions

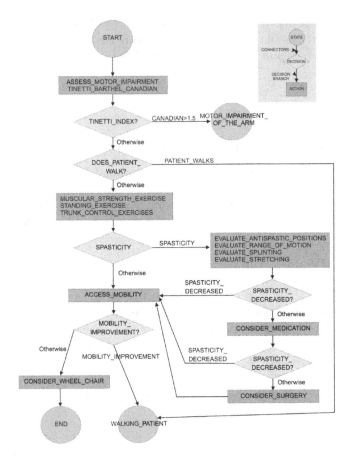

Fig. 2. Example of SDA* Formal intervention plan of *post-stroke rehabilitation*

at the same time. The recommendations made by several FIPs must be taken into account to construct an IIP that indicates the personalised care actions to be applied on a specific patient.

The creation and management of IIPs follow a complex procedure which is controlled by a multi-disciplinary team called *Evaluation Unit* (EU) (see Figure 3). An EU includes four actors: the physician in charge of the Home Care unit (PC), a family doctor (FD), a social worker (SW) and the head nurse (HN). All these actors have their own persistent agents within the platform, while the EU has not (the EU is a logical grouping needed for service management). The EU selects the professional to cover the role of Case Management, who manages the overall accomplishment of the IIP (usually the HN). The EU is a temporary team and only constitutes in the circumstances of the first assessment of a HCP and of the periodical re-evaluations. It is possible that the same EU takes care of different patients.

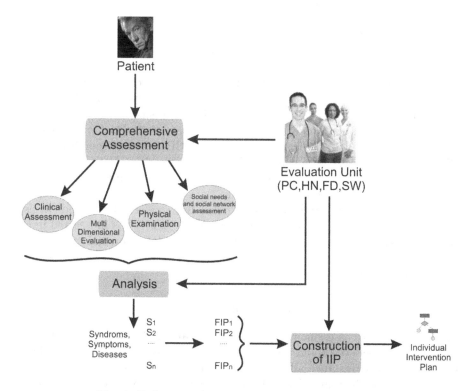

Fig. 3. Definition of an Individual Intervention Plan

A typical scenario of HC provision involving the life-cycle of an EU can be presented by the following service sequence: *i*) HC Request, *ii*) HCP Admission, *iii*) EU Constitution, *iv*) EU-HCP Binding, *v*) Comprehensive Assessment, *vi*) Planning and executing an IIP and *vii*) EU Dissolution.

HC Request permits the system to know that a person needs care (FD demands HC for a HCP) and HCP Admission allows the K4Care model to know which are the patients involved in the HC service and which are not (the admission is done by the PC and HN, under FD request).

The EU takes form in the platform through the EU Constitution service. At the moment of the EU's constitution three of the four necessary actors are already well-defined and unambiguous: the FD (he requested the HC for the patient), the PC (it is unique in a HC structure) and the HN (it is also unique as the PC). The only actor to be assigned (by an Actor Assignment action) is the social worker, and the HN is responsible for the achievement of this task. Different strategic and/or decisional policies for human resource management can be applied at this moment by the HN (actor availability, work load, closeness to patient).

On the basis of the previous considerations, the HN finally sends a message to all the elected actors as a confirmation of their assignment to a new EU. Trace of the EU constitution is kept both in the local memory of each member agent and in a permanent storage. This prevents from data loss in the case of platform crashes. Once an EU has been constituted, it must be officially bound to the patient whose medical needs it was created for. This is accomplished by the EU-HCP Binding service.

The first step in the patient's care is the performance of a comprehensive assessment (CA), which includes a multi-dimensional evaluation (filling a set of internationally standardised scales), a clinical assessment, a physical examination and a social needs and social network assessment. Thus, the CA is the service devoted to detect the whole series of HCPs diseases, conditions, and difficulties, both from the medical and social perspectives. It is performed at admission, at periodical (or end-treatment) re-evaluation times, and in case of emerging peculiarities during the follow-up.

Once the CA has terminated, the EU still has model validity and persists within the platform. It is ready for the definition of an IIP, as illustrated in the following section. The successful completion of an IIP definition and execution for a patient triggers the automatic dissolution of the EU, by invoking the execution of the service named EU Dissolution.

3.3 IIP Semi-supervised Definition

The comprehensive assessment represents a fundamental process on the basis of which the members of the EU are enabled to reason and react on the patient's state and conditions. Once all the results of this service are available, the EU members analyse them in order to determine syndromes, symptoms and diseases of the patient. Basically, any of these entities is associated with a FIP, which is a very general intervention plan. A FIP does not represent the condition's evolution for the patient following the intervention plan and is not directly applicable to any real treatment or intervention. This is due to the uniqueness of each patient's physical conditions and medical history. It is a usual practice, on the contrary, to recognize at the same time several diseases to treat, which can lead to a very complex and articulated assessment.

The platform automatically retrieves all FIPs concerning and dealing with the patient conditions. The EU members can then use the SDA* graphical editor, which is embedded in the web interface, to combine and personalise the relevant sections of these FIPs in order to build the specific IIP for the particular patient at issue (see Figure 4).

Trying to automate the union process of different FIPs, in order to provide the HC staff with a supporting tool in the definition of a new IIP, is not a trivial task and implies, among other aspects, an accurate feasibility analysis of the objective. This is mainly due to the complexity and heterogeneity of the physical parameters and the HC domain variables. Anyway, at the moment of writing this paper, a supporting mechanism is being analysed and conceived in K4Care, to be integrated in the SDA* editor. Some preliminary ideas and guidelines are reported in the following sections.

In the rest of the section, the procedure to create IIPs will be explained step-by-step. The first stage consists in the *creation of a model*. Simplifying, it is

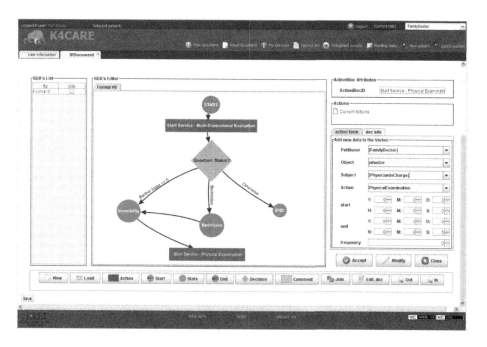

Fig. 4. Visual editor of Individual Intervention Plans

necessary to: verify the preconditions for the integration of FIPs, and verify the compatibility of the different drugs prescribed in the FIPs of interest, whose prescription and use in the different FIPs may be in conflict. When the model is created, a formal verification of the model avoiding non-deterministic sequences of actions is required. Finally, manual revisions should be performed in order to *customize the IIP* for the current patient.

IIP's creation model. To facilitate our investigations, we can consider a FIP as a black-box from the point of view of its provision steps (provision logic). What is important in this stage is the FIP's capability to interface with and to plug into the outer level of its execution (FIP interfaces are represented with dotted rectangles in Figure 5). In practice, a FIP's interface consists of an Entry Points (EP) set. EPs are essential because they represent an execution's possible starting point in the provision of the FIP.

In an SDA* model, a patient condition contains all the variables observed for the patient in a particular moment (*i.e.*, signs, symptoms, antecedents, taken drugs, secondary diseases, test results, etc.). State variables are contained in the EPs of a FIP and their value is expressed by logical conditions (example: PRESSURE<140 and WEIGHT>75), which are at the basis of the verification of preconditions in the FIPs integration's process. We can summarize the procedural steps of the integration process as follows:

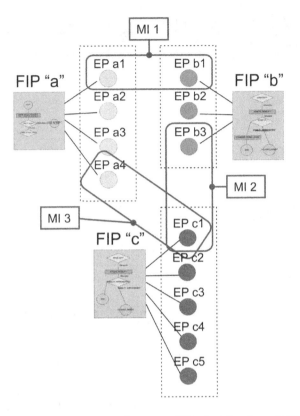

Fig. 5. Interaction example among EPs deriving from 3 FIPs

1) Analyse all sets of conditions contained in the EPs of all FIPs and list the state variables.
2) Consider the possible intersections of EPs in all FIPs, that is, the set of state variables that are shared among EPs of different FIPs.
3) If the intersection is *empty*, the FIPs can be considered as independent of each other, and the new IIP can be automatically defined as the union of different minor IIPs (each of them deriving from its corresponding FIP), to be personalised, merged and completed by medical staff, as reported at the end of this section.
4) If the state variables intersection is *not empty*, then we can assert that some Model Interaction - MI - has occurred among the original FIPs (rounded corner rectangles in Figure 5). Hereby, we define a MI as a couple of EPs of two different FIPs that share at least one variable (1 or more than 1 variable).

At this point, the verification of conditions associated to state variables in common must be taken into account. For each MI, we can formalize the following additional steps:

5) If all conditions relating to those variables shared between 2 different EPs in 2 different FIPs are compatible[1], then both original EPs causing the MI can be replaced by a single new EP. This new EP will be non-deterministically connected (*stars* in Figure 6) to both FIPs in which the original EPs from the MI were located. The following example, together with the representation in Figure 5 and Figure 6, aims at clarifying the point of the previous statements.

Example 1: Consider the following two EPs:

EP a1: (SUGAR>40) and (50<PRESSURE<100) and (IRON<20);
EP b1: (SUGAR>20) and (70<PRESSURE<90) and (AGE>65);

We can define the new entry point as follows (circles coloured in light cyan in Figure 6):

NewEP = EP n-a1b1: (SUGAR>40) and (70<PRESSURE<90) and (IRON<20) and (AGE>65).

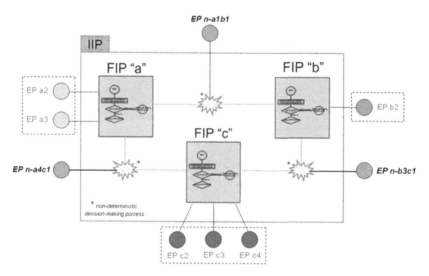

Fig. 6. IIP definition model in case of compatible conditions in all FIPs entry-points

In other words, for shared variables we take the *most constraining condition*, and we keep the other original conditions from the two EPs as they are. It is easy to verify that a patient entering the IIP through the new Entry Point (*EP n-a1b1* in Figure 6) could have also entered it through *EP a1* or *EP b1* (Figure 5). The opposite on the contrary, would not hold, since a patient with sugar 50, pressure 60 and iron 15 could go through *EP a1* of FIP "*a*", but not through *EP n-a1b1* of the IIP. Entry points not involved in any MI will be naturally keept in the new IIP (rounds coloured in yellow, green and blue in Figure 5 and Figure 6).

---

[1] Note that "compatible conditions" only refers to the fact that there is at least 1 (one) value that satisfies both conditions.

Figure 6, furthermore, shows two additional Model Interactions occurring among the original FIPs: one between FIP "b" and FIP "c" (leading to the EP n-b3c1) and the other between FIP "a" and FIP "c" (leading to the EP n-a4c1).

6) If there is at least 1 variable leading to incompatible conditions, the involved EPs cannot be merged and in the resulting IIP both EPs will be kept as they were in the original FIPs (red-coloured circles in Figure 7).
This case is illustrated by the following example.

Example 2: Let us consider the following EPs:

EP a1: (SUGAR>40) and (50<PRESSURE<100) and (IRON<20);
EP b1: (SUGAR<15) and (PRESSURE>120) and (AGE>65);

In this case, the shared conditions on sugar and pressure are incompatible, therefore, as depicted in Figure 7, the EPs are not merged.

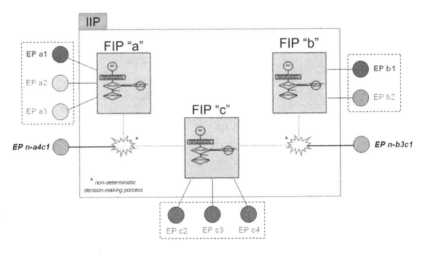

Fig. 7. IIP definition model in case of incompatible conditions between *EP a1* and *EP b1*

Once the FIPs integration's verification phase terminates, integration incompatibility reasons are provided to practitioners as a possible decision-making support base towards a manual arrangement of the intended IIP's definition.

Towards a deterministic evaluation of new entry point branches. As reported in the previous paragraph, the introduction of a new entry point comes with a non-deterministically connected branch to both original FIPs (non-deterministic branches are represented as a *star* in Figure 6). The reason for such a flow uncertainty lies on the initial impossibility to univocally decide which is the starting FIP the new EP should be connected to.

Despite the objective decisional complexity, some kind of deterministic metrics (even though elementary) could be preliminarily introduced, in order to automate the continuation of the IIP provision. We proceed to explain this conceptualization with the following example.

Example 3: Let us briefly recall the EPs from the *Example 1*:

EP a1: (SUGAR>40) and (50<PRESSURE<100) and (IRON<20);
EP b1: (SUGAR>20) and (70<PRESSURE<90) and (AGE>65);
NewEP = EP n-a1b1: (SUGAR>40) and (70<PRESSURE<90) and (IRON <20) and (AGE>65).

Now consider the following possible patient state conditions, before entering the *EP n-a1b1*. Let's assume that practitioners can consider a variable as *not relevant* when its measurement is not mandatory for a specific assessment:

PSC1: (SUGAR=50) and (PRESSURE=75) and (IRON=10) and (AGE = Not-Relevant);
PSC2: (SUGAR=60) and (PRESSURE=75) and (IRON =Not-Relevant) and (AGE=80);
PSC3: (SUGAR=60) and (PRESSURE=80) and (IRON =Not-Relevant) and (AGE =Not-Relevant);
PSC4: (SUGAR=60) and (PRESSURE=85) and (IRON=15) and (AGE=80);

The evaluation, case by case, of the branch automatically chosen when entering the *EP n-a1b1* (decision-making processes in Figure 6) can be summarized as follows:

1) *Case PSC 1*: all relevant variables (and conditions) involved belong to the *EP a1* of FIP "*a*", while the last variable, the age, though present and belonging to the *EP b1* of FIP "*b*" is not relevant for the evaluation of the patient condition. From here, we can assume that the IIP execution can automatically continue towards the branch of FIP "*a*".

2) *Case PSC 2*: all relevant variables (and conditions) involved belong to the EP b1 of FIP "b", while the last variable, the iron, though present and belonging to the *EP a1* of FIP "*a*" is not relevant for the evaluation of the patient condition. From here, we can assume that the IIP execution can automatically continue towards the branch of FIP "*b*".

3) *Case PSC 3*: this is not a straightforward case, because all relevant variables describing the patient's condition are exactly contained in both FIP "*a*" (*EP a1*) and FIP "*b*" (*EP b1*).

In this case we can determine, for example, how close the measured state's value for a variable is to the boundaries of the condition containing the same variable in the different EPs.
In other words, we can measure to which degree the variable's value is contained in the condition range. The closer the value to the middle point of a range (presence of lower and upper boundary), the more the condition can be considered as satisfied. In case of a single-boundary condition, we can

take into account how far the variable value is from the boundary itself. The previous evaluation must be repeated for each variable and, finally, the majority of verified conditions will lead to a FIP instead to another. Recalling here the actual case, we have:

- SUGAR=60, which better (more largely) satisfies the condition of *EP b1*: (SUGAR>20) than *EP a1*: (SUGAR>40). One point for FIP "b".
- PRESSURE=80, which better satisfies the condition of *EP b1*: (70<PRESSURE<90) than *EP a1*: (50<PRESSURE<100), since *EP b1*'s middle point is 80 while *EP a1*'s is 75. Another point for FIP "b".

Summarizing, this particular case would automatically address FIP "b" as the continuation of the IIP provision.

4) *Case PSC 4*: this case leads back to the non-deterministic approach previously mentioned. Further information must be taken into account here in order to have decisional capability. For example, important guidelines are implicitly provided by the nature of the FIPs: though they can have variables in common, their provision logic (care treatment) can address completely different diseases. From here, we can agree the necessity to improve the representation of the care provision knowledge.

IIP definition's final customization steps. The IIP resulting from the automated definition phase is not a complete and executable IIP in the platform: personalised drug quantifications, time constraints in the execution of actions, non-deterministic events management, and other aspects, must be explicitly added to the model, as well. Once the final IIP is ready, it is saved in the Electronic Health Record of the patient. Thus, the medical team does not have to build the IIP from scratch, and can take into account the international recommendations in the treatment of the patient's conditions, composing a fully personalised and accurate care plan for him. The IIP usually contains follow-up actions in which the state of the patient is checked. If the evolution of the patient through his customized plan follows an undesirable course, the EU could consider changing or even cancelling the IIP.

3.4   Ontology-Driven Execution of an IIP

An IIP contains diverse elements, such as expressions included in states, logical conditions included in decisions, and actions to be performed by actors. All these data are formatted accurately in order to know exactly what anyone wants (*know-what*) and who should perform it (*know-how*). This knowledge is represented in two ontologies that are consulted in every step of the execution: the APO and the CPO (partial views of these ontologies are shown in Figure 9). Medical experts, with the help of knowledge engineers, have manually constructed them. The APO includes the structural knowledge of a medical organization. Besides the permissions of each actor, it is also possible to find there the different documents used in the K4Care platform (and their read/write permissions), the procedures, their associated SDAs, and the list of actions (related with the actors who can do these actions). The CPO represents the relationship between the different signs and symptoms that a patient can present with the different diseases and syndromes (both coupled with their interventions) that the patient could have, linked with the standard codes used to identify them [2].

The main actors involved in the execution of an IIP are the Head Nurse's Actor Agent, which controls the step-by-step execution of the whole IIP, and the SDA-Executor Agent (SDA-E), which collects the required data from the EHR and forwards the successive elements of the IIP to the AA of the HN [9]. In the first iteration of the execution process the SDA-E agent should look for all entry points (states without preconditions). The EU should determine the exact entry point in each case. After that, the AA of the HN will receive the information on the next step of the IIP (state, decision or action).

In the case of a *state*, the expressions contained in it should be forwarded to the AA of the EU members so that they can check whether the patient is indeed in that clinical situation.

When a *decision* is found, its logical expressions, which contain decision variables and conditions, should be evaluated (*e.g.,* TINETTI_INDEX>1.5). The location of these variables in the documents is located through the CPO ontology via the DAL. The DAL provides transparency to the agents and retrieves all this information automatically.

Finally, the last element that the SDA-E agent may send to the AA of the HN is an *action*[2]. Figure 8 shows the interactions between all partners involved in this case [9]. Concretely, the SDA-E sends a task, containing an action identifier and a type of actor as subject (steps 2-4). If the required actor belongs to the EU, the assignment is direct to this actor. Otherwise, the HN has to select a concrete person to perform the action. The platform provides up-to-date information about all required actors to the Head Nurse that can quickly select the name of the addressed one. When this specific person logs into the system, he will find the request in the list of pending actions. The HN should identify which document has to be filled according to the action to be performed (step 5). After selecting this action, a message will be sent by his *Gateway Agent* to his *Actor Agent* (steps 6-8). Then, the *Actor Agent* will retrieve, using the appropriate method of the DAL, the empty document to be filled out in order to reflect the performance - in the real world - of the action (step 9). The Actor Agent sends the document to the web interface (through its associated Gateway Agent), and the practitioner fills it out consequently[3] (step 10). When the document is completed (step 11), the Actor Agent stores it the EHR of the patient and sends a confirmation message to the Actor Agent of the HN to indicate that the action has been successfully performed (step 12). After that, the Actor Agent of the

---

[2] The K4Care Model includes 84 specific care actions that may be performed by 10 different types of actors.

[3] There are 43 documents defined in the K4Care Model (some documents are general enough to be usable to reflect the performance of different actions).

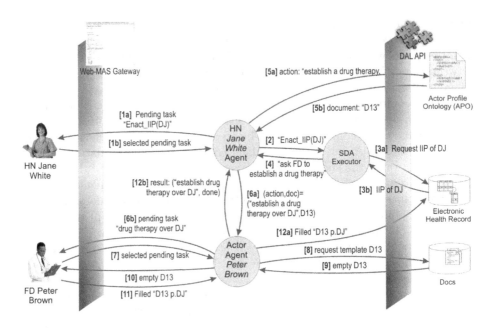

Fig. 8. Description of the action execution procedure

HN can request to the SDA-E the next element in the IIP execution (which may be a state, decision or action).

The APO indicates which actions are allowed for each kind of actor (see Figure 9(a)). Sometimes the actor assignment is simple, because the patient and the type of agent directly imply a given agent instance. If the action has to be taken by a Physician in Charge, a Family Doctor, a Head Nurse or a Social Worker, then the assignment is made automatically to the corresponding person in the EU, for a given patient. Otherwise (*e.g.,* if a nurse or a specialist physician is needed) the HN has to select a concrete person to perform the action. The platform provides a pop-up menu in the GUI with the actors currently available so that the Head Nurse can quickly select the name of the addressed one. The concrete list of actors is dynamically compiled by the system according to the Actor Agents currently running, to their associated type of actor and, in consequence, to the list of actions they can perform (information picked from APO). It is important to note that, as the system supports the personalisation of an actor's profile, this should be taken into consideration when assigning the action or proposing the actor as a performer. Concretely, an actor may decide, at runtime, not to perform some kind of actions. This request is processed by means of the personalisation of an individual copy of the Actor Profile Ontology. Once the changes are validated by the Physician in Charge, the new personalised profile will be taken into consideration immediately. From the point of view of the action assignment, this leads to the fact that the system automatically requests each Actor Agent whether the pending action may or may not be performed in that specific moment by the concrete actor.

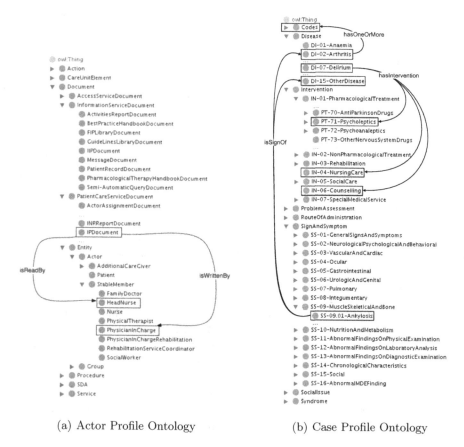

(a) Actor Profile Ontology   (b) Case Profile Ontology

Fig. 9. Partial view of the ontologies designed in the K4Care Project

4   Conclusions and Further Work

Chronic and disabled patients usually require long-term treatments under the continuous supervision of experts. These treatments represent real challenges for National Health Systems, requiring a differentiation of medical assistance in health centres from ubiquitous assistance (HC Model).

This paper has presented the creation and execution of personalised intervention plans *adapted* to each patient. Particular importance has been given to the key aspect of HC knowledge and guidelines representation (SDA*) which are at the basis of an IIP's management.

The execution of IIPs is a very complex procedure, which is controlled by the EU multi-disciplinary team, whose decision capability relies on the results of the CA service.

When all the results of the Comprehensive Assessment are available, the platform could use the medical knowledge stored in the CPO to help the EU to determine the syndromes, symptoms and diseases of the patient (*e.g.*, if the blood pressure is above a certain level, it could suggest the hypertension disease automatically). This is a promising line of future work.

It is also our intention to continue working on different ways in which the K4Care Platform may support the EU members in the complex process of merging several FIPs to obtain an IIP (right now, this process is completely manual). For instance, if the CPO contains information about drugs that cannot be taken simultaneously, the platform can check in the IIP that there is not any branch that includes two incompatible drugs.

In order to show the immediate applicability of the model, a more refined version of the platform will be deployed for the community of the town of Pollenza (Italy). The test deployment will involve the entire Home Care facility, General Practitioners, the Municipality, social assistants and citizens' representatives. The number of patients in this assessment is expected to be around 100. Finally, a professional software development company is being hired at this moment to develop a final version of the platform and perform an exhaustive technical testing (concurrently to the medical evaluation) in order to obtain a robust and reliable software product as the final project's result.

Acknowledgements. This work has been funded by the European IST project K4Care: Knowledge Based Home Care eServices for an Ageing Europe (IST-2004-026968). The authors would like to acknowledge the work of the other members of the K4Care consortium, especially the medical partners, led by Dr. Fabio Campana, and Dr. David Riaño (project co-ordinator and designer of the SDA* formalism).

The authors are solely responsible for its content. It does not represent the opinion of the European Community and the Community is not responsible for any use that might be made of the information contained herein.

References

1. Wyatt, J., Sullivan, F.: eHealth and the future: Promise or Peril? British Medical Journal 331, 1391–1393 (2007)
2. Campana, F., Moreno, A., Riaño, D., Varga, L.: K4Care: Knowledge-Based Homecare e-Services for an Ageing Europe. In: Agent Technology and e-Health. Whitestein Series, pp. 95–115. Birkhäuser, Basel (2008)
3. AGS: The American Geriatrics Society (2007), http://www.americangeriatrics.org/ (last visit, 2008/12/04)
4. NGC: National Guideline Clearinghouse (2007), http://www.guideline.gov (last visit, 2008/12/04)
5. Durso, S.: Using Clinical Guidelines Designed for Older Adults With Diabetes Mellitus and Complex Health Status. Journal of the American Medical Association 295, 1935–1940 (2006)
6. Elkin, P.L., Peleg, M., Lacson, R., Bernstam, E., Tu, S.W., Boxwala, A., Greenes, R.A., Shortliffe, E.H.: Toward Standardization of Electronic Guideline Representation. MD Computing 17, 39–44 (2001)
7. Isern, D., Moreno, A.: Computer-based Execution of Clinical Guidelines: A Review. International Journal of Medical Informatics 77(12), 787–808 (2008)

8. Cabana, M.D., Rand, C.S., Powe, N.R., Wu, A.W., Wilson, M.H., Abboud, P., Rubin, H.: Why don't physicians follow clinical practice guidelines? A framework for improvement. Journal of the American Medical Informatics Association 282, 1458–1466 (1999)

9. Isern, D., Millan, M., Moreno, A., Pedone, G., Varga, L.Z.: Agent-based execution of Individual Intervention Plans. In: Moreno, A., Annichiaricho, R., Cortés, U. (eds.) Workshop Agents Applied in Health Care, collocated with AAMAS 2008, Estoril, Portugal, pp. 31–40 (2008)

10. Riaño, D.: The SDA* Model: A Set Theory Approach. In: Twentieth IEEE International Symposium on Computer-Based Medical Systems, CBMS 2007, Maribor, Slovenia, pp. 563–568. IEEE Press, Los Alamitos (2007)

11. McGuinness, D., van Harmelen, F.: OWL Web Ontology Language (2004), http://www.w3.org/TR/owl-features/ (last access, 2008/12/04)

12. Batet, M., Gibert, K., Valls, A.: The Data Abstraction Layer as Knowledge Provider for a Medical Multi-agent System. In: Riaño, D. (ed.) K4CARE 2007. LNCS (LNAI), vol. 4924, pp. 87–100. Springer, Heidelberg (2008)

13. Tinetti, M.: Preventing Falls in Elderly Persons. The New England Journal of Medicine and Health 48, 42–49 (2003)

Electronic Health Record as a Knowledge Management Tool in the Scope of Health

Miguel Angel Montero[1] and Susana Prado[2]

[1] IECI, Travesía Costa Brava, 4
28034 Madrid, Spain
[2] IECI, Gran Via de les Corts Catalanes 613, 6°
08007 Barcelona, Spain
{miguel_monterom,susana_prado}@ieci.es

Abstract. The concept of knowledge management, introduced by Peter Drucker at the end of the 90s and promoted by figures like Gary Hamel and Coimbatore K Prahalad, is based on the idea that companies must throw themselves into the knowledge of their professionals in order to obtain a competitive advantage in front of their direct competitors, either on costs or on differentiation, and to reach the higher value for the client. In the last years, this conception, linked to the post-capitalist society, is starting to be applied in the scope of health.

This requirement is translated on the fact that managers of health centers, regardless of if they are private or public, direct the information systems towards their healthcare professionals, essentially physicians and nurses, since they are the ones who have the knowledge to generate more added value in the patient, which is the ultimate reason of being of any healthcare organization. Electronic Healthcare Record Systems can be considered a subsystem of knowledge management.

This paper summarizes the pillars to hold up the knowledge generated within a healthcare organization using the Electronic Health Record as the core element and engine for the more advanced Knowledge Management Systems. The paper compares the most extended market solutions of the healthcare environment in the Spanish national territory.

1 Introduction

Informática El Corte Inglés (IECI[1]) constitutes the matrix of the area of companies related to Information and Communication Technologies of the El Corte Inglés Group. The supply of IECI has evolved from the distribution of computer products and solutions, initial aim of its constitution, to the consultancy and integration of electronic business systems and solutions. Both products as well as integrated third parties products and solutions are used in order to offer the most competitive technological solutions of the market with a complete vision both from a functional level and from a projects lifecycle level (i.e., consultancy, analysis, definition, installation and maintenance). More than 2400 professionals are working in IECI combining the best business strategy

---

[1] http:// www.ieci.es

D. Riaño (Ed.): K4HelP 2008, LNAI 5626, pp. 152–166, 2009.

profiles with the highest levels on technological knowledge. In 1997, IECI created a group specialized in healthcare environments and management, in 1998 it carried on the whole Plan of Technological Renovation of the National Institute of Health (INSA-LUD[2]) of the Ministry of Health of Spain. In 1999, IECI bet on the current technological solutions installing SIAH in the hospitals of the Public Catalan Health Systems (ICS) and HP-HIS in the hospitals of INSALUD. In 2000, IECI carried out a consultancy on the information systems of the new Sant Pau Hospital in Barcelona, installing afterwards SAP in all its areas (i.e., Economic and financial, Logistic, Human Resources, Healthcare and Clinical). Moreover it has developed consultancy and development projects in the Councils of Health of the different Spanish Autonomous Regions and in several hospitals: Delegation of Barcelona, ICS, Catalan Institute of Oncology, Government of Aragon, Principality of Asturias, Baliaric Islands Health System, Community of Madrid, Canary Islands Health System, Navarrese Service of Health, Andalusian Service of Health, Service of Health of Castilla-La Mancha, Badalona Municipal Hospital, etc.

With this experience, IECI is, nowadays, a great expert on the healthcare configuration in the Spanish territory, a national point of reference in systems for the computerization of health, an expert on the health computer systems that exist in the market and an expert on several installation experiences of these systems at the Spanish, European and world-wide levels.

This knowledge and experience allows us to state the increasing interest and necessity of healthcare organizations to have tools to allow them to manage the clinical information in a suitable way. We have witnessed how healthcare professionals are demanding management computer-based tools for the development of their professional activity. Similarly our experience justifies the IECI capability to carry out a comparative analysis of the different market solutions for knowledge management in the scope of health, whose results are presented in this article.

1.1 Data as a Fundamental Pillar for Knowledge Management

In the context of healthcare, data are primary aspects. These data are transformed into information when they indicate an evaluation or an action as it could be the diagnosis of arterial hypertension. For that reason the information in the scope of health is conceived as the pillar on which clinical decisions are based. Data can be of different types, like for example, administrative, basic healthcare or clinical complexes (e.g., diagnoses, procedures, evolutions, comparatives, assessments, etc.).

1.2 Knowledge Management in Health

In health care organizations, only on the basis of a well structured and managed information of healthcare and clinics it is possible to introduce knowledge management in a coherent way.

Sánchez-Calas, when referring the concepts of information and knowledge in the scope of health [8], differentiates between explicit, tacit, and implicit information, according to the source from where they come. Thus, *explicit information* is the result of scientific research in the biomedical sector, but also of the evaluation of healthcare

---

[2] Nowadays National Institute on Health Management (www.ingesa.msc.es).

services and assistance. *Tacit information* is the result of healthcare professionals experience in their clinical practice, and *implicit information* is contained in clinical files.

Similarly, *explicit knowledge* is acquired by means of documentary sources, both internal and external. *Tacit knowledge* is the one present in people as a result of their experience and *implicit knowledge* results from the working practice of healthcare professionals.

The main knowledge sources in the area of health are healthcare professionals who *own the tacit knowledge intrinsically*. This sort of knowledge is the one with a greatest value in the knowledge management process. Obviously there are other significant sources of knowledge [20] that are being used and that corroborate the idea that health is a suitable field for knowledge management. Some of these are:

- **Traditional sources:** They are typical of healthcare centers, such as conventional clinical records, methodological documents, statistical summaries, scientific events, accounting and budget reports, bibliographical information in paper support, internal reports, or tutorials, among others.
- **Outsourcing:** All those that include aspects related to healthcare centers but that are not part of them, such as Benchmarking, use of specialized consultants, clients, suppliers or extern reporters.
- **The Internet:** It offers certain facilities as a valid source of intelligence for decision making by the doctors, so it contributes both to the increase of the communication and to the development of research and, in general, to a more effective knowledge management in health.

  Among other things, the Internet allows accessing medical databases, systematic revisions, clinical practice guidelines, journal articles, professional associations of the scope of health, congresses presentations and reports, reference documents as dictionaries, catalogs, etc.; as well as interchanging experiences and knowledge with other professionals, inside or outside of the country, which is boosted by new means as social networks.

- **Team Work:** This source of knowledge represents the most important component of the knowledge management process in healthcare, where it was applied from long before it was used in other sectors. Team work is required for discussing cases, surgical operations and interconsultations; all of them processes in which knowledge is generated and exchanged.

  In addition, team work supposes a particular and customized form of knowledge, since it allows transmitting tacit knowledge and experiences, beyond the content of the documents of health records, since team work is based on the experiences and the intuition of professionals. This information which is stored and accumulated in time can be exchanged with the obtained results to protocolize and to establish best professionals practices.

So, Integrated Managament Systems in health are those that allow combining all the services and departments of an organization or healthcare system, with the electronic health record taken as core element with a clear orientation to patient and professional needs. These systems contributate to strengthen team work and to increase the generation of collective knowledge.

1.3 The Electronic Health Record as a Subsystem of Knowledge Management in Health

An organization owns explicit knowledge contained in the documents that are part of their assets. These documents are of great value if they contain the experience in the form of information, and if this information is transformed into usable knowledge when the staff of the institution uses it.

In healthcare organizations, Healthcare Records (HR) are knowledge management systems that integrate all the information associated to the patient. HR enables the organization to solve the problems of the patients. Moreover, HR is well adapted to the concepts of data and tacit and explicit information and knowledge. We must notice that HR is a repository of healthcare knowledge that can be in the memory of professionals, in paper documents or in electronic records.

On the basis of all this in the healthcare context, we can affirm that HR is a subsystem of Knowledge Management.

In this article we will talk about Health Record in electronic format or Electronic Health Record (EHR) as: *"The health record in which the information and the documents are in computer support and therefore can be consulted, processed, transmitted and presented using information and communications technologies"*.

The EHR concept comes from the original Computer-Based Patient Record (CPR) [13, 14], for which four generations are distinguished whose differences are exposed in table 1.

Table 1. CPR Generation KM Criteria

Generation	Criteria
1st	None.
2nd	• Stand alone capabilities vs. integrated capabilities
3rd	• Capability of the CPR to interact with KM content • Integration of KM information with clinical decision logic • Has a knowledge base for clinicians to store medical literature references • Include means to search the KM repository for complete retrieval on medical concepts
4th	• Full support for searching the KM content • Capable of including vendor- and client-defined metadata • Uses a formal knowledge representation language • Perform consistency checking on the information • Support care management protocols • Has tools to accommodate multiple concurrent protocols for a single patient • Supports "context-aware" clinical queries (e.g. literature searches, potential clinical trials or alternative therapies). • Capability of including institution policies and procedures related to clinical care. • KM content is equipped to support life cycle management.

2 Healthcare Contexts in Which Knowledge Management Technologies Can Be Interesting

Although EHR can provide legal support, or allow the development of population studies that constitute another type of knowledge with a purpose like, for example, resource allocation, the basic objective of EHR is to integrate knowledge on the health of patients. Therefore, the contexts of healthcare in which EHR for knowledge management can have an outstanding interest include *healthcare planning* (i.e., Public Healthcare Systems), *management* (i.e., healthcare services, hospitals or primary care areas) and the *support to professionals* (i.e., medical, surgical and nursing services).

In all the stated levels, the EHR is a powerful tool for knowledge management, but each level will obtain different benefits according to the characteristics of its activity.

3 Tools for an Electronic Health Record

There exist multiple tools and systems to manage an EHR. We should ask why the current viability of these systems has not been observed in another previous time.

Although the technological development is an important factor, currently it is necessary to consider other factors as, for example, the reasons of the different institutions and services of health (either at local, regional, national, European and worldwide levels) to consider EHR the most relevant project within the scope of e-Health [19, 21, 22, 23].

In the settings where healthcare knowledge is shared by means of EHR, the contribution of knowledge to the system implies a user cost that is proportional to the effort of elaborating new information to share. Moreover, the more users collaborate in the system, the higher is the immediate benefit, but the higher the global effort may be [24].

Nowadays, there is an unanimous belief on the benefits of having an EHR which is shared either by managers [26], by healthcare professionals [25, 27], by citizens and by patients. Therefore in theory, nowadays the EHR does not have the risk that is observed in other systems for knowledge management in which users can adopt attitudes of starving the system of information and only expecting the benefit of their exploitation. These undesired attitudes affect both the quality and the amount of the contents and, indirectly, the ability of generating the awaited knowledge and results.

There are several tools for the elaboration and management of EHR, but we will concentrate in the ones most used: SAP [9], SELENE [15], HCIS [16] and CERNER [17].

3.1 SAP: IS-H/IS-H*MED

Figure 1 depicts the solution proposed by SAP for healthcare [10, 11]. It is based on four axes: strategic planning, resources and logistic, treatment and relation with the patient. In this background, SAP offers several services that are organized in strategic services, planning services, collaborative services, patient management services, and enterprise management services. At the product level, SAP implements this solution by means of Industry Solution Hospital (IS-H) and Industry Solution Clinical Hospital System (IS-H*MED) [9]. IS-H is an improvement of the standard R/3 System specialized in the hospital management and with elements for patient management, both medical and nursing documentary management and accounting management. It also supports the communication and transfer of information inside and outside the hospital. IS-H*MED extends the functionality of IS-H with scheduling and process control.

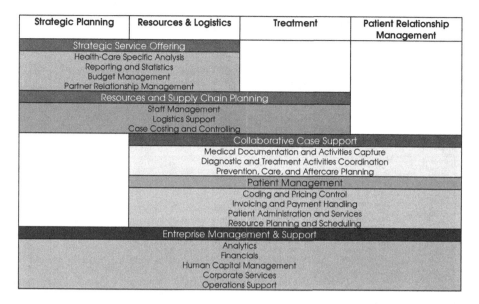

Strategic Planning	Resources & Logistics	Treatment	Patient Relationship Management
Strategic Service Offering			
Health-Care Specific Analysis Reporting and Statistics Budget Management Partner Relationship Management			
Resources and Supply Chain Planning			
Staff Management Logistics Support Case Costing and Controlling			
	Collaborative Case Support		
	Medical Documentation and Activities Capture Diagnostic and Treatment Activities Coordination Prevention, Care, and Aftercare Planning		
	Patient Management		
	Coding and Pricing Control Invoicing and Payment Handling Patient Administration and Services Resource Planning and Scheduling		
Entreprise Management & Support			
Analytics Financials Human Capital Management Corporate Services Operations Support			

Fig. 1. SAP solution for knowledge management in health

In general among other things, SAP proposal for hospital knowledge management by means of EHR allows creating an integrated EHR of the patient, using the EHR in a customized way according to the different departments or services of a healthcare center and using interoperability standards for the healthcare environment.

IS-H*MED has been established in more than 272 hospitals world-wide [12]. The advantages of this solution could be summarized in:

- Core component of one of the most innovative clinical solutions of Europe.
- Interfaces that are learned to handle with facility (see figure 2).
- Simple use and easy navigation applicable to all the clinical scopes.
- Effective scheduling thanks to its high integration.
- Centralized agendas for patients, staff and departments.
- Transparent communication of requests and results.
- Powerful multimedia documentation with digital dictation and voice recognition.

The main advantage of the tool is the integration with the rest of information subsystems in healthcare: patient management, subsystems of economic and logistic management and human resources.

Some disadvantages of SAP are expressed as the following features: excessively management-oriented tool with consolidated ergonomics in industrial environments (not clinical) requiring parametrization efforts oriented to improve the interface with the healthcare professional; this configuration-implementation is not very intuitive and it requires technology experts to fit and to facilitate the access to the information.

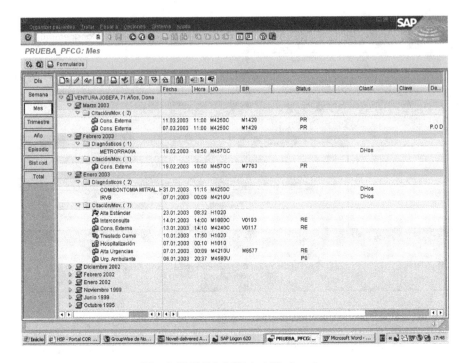

Fig. 2. IS-H*Med Clinical Workstation

3.2 SIEMENS: SELENE

The Clinical Station of SIEMENS Spain is SELENE SIEMENS [15]. It involves the management of all the clinical information of the patient (defining in this way the concept of EHR) and making all the clinical staff to participate of this information in their professional activities.

From a unique identification of the patient, this Clinical Station displays a perspective of global healthcare information of the patient which is organized in **Healthcare Processes** that reflect each one of the relations of the patient with the healthcare system. This relation takes place in different places and moments (e.g., primary care, hospital, center of specialty, ambulance, etc.) and it is necessary to equip it with unicity, for example personalizing the assistance of a patient who enters with appendicitis.

The possibilities of configuration and personalization are brough to the level of end-users, making the adaptation of the tool to any work stream possible and tailoring work environments adapted to each user in each one of the scopes in which he or she carries out his or her activity.

The scheme in figure 3 depicts the basic blocks that compose SELENE for medical knowledge management based on EHR [28]. It establishes services for the clinical professional to introduce data and for management of administrative tasks, healthcare activities and requests. So, it is based on the concepts of clinical route established by the center and workflow.

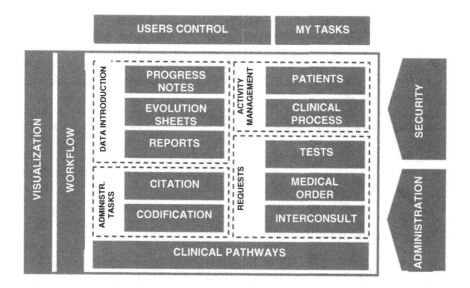

Fig. 3. SIEMENS Spain solution for knowledge management in health

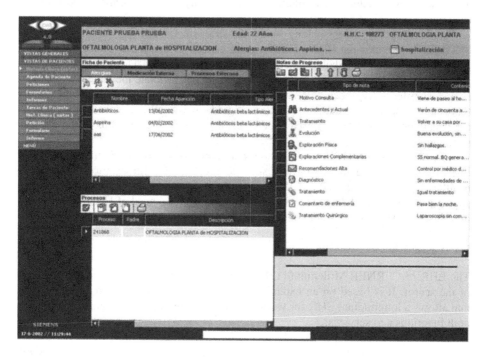

Fig. 4. SELENE Clinical Station

It is important to stand out that in 2007 SIEMENS acquired Gesellschaft fuer Systemforschung und Dienstleistungen im Gesundheitswesen mbH (GSD), the company that developed IS-H*MED. This acquisition is giving rise to a compilation of functional contents of IS-H*MED and Soarian solutions that will probably make SIEMENS change the strategy with respect to SELENE, starting the replacement of SELENE in their clients.

SELENE has a clear orientation to clinical management (see figure 4), being this its main assets, however the integration with the patient management is one of its weaknesses.

Unlike the SAP solution, SELENE is only present in Spain, since the SIEMENES strategy for the rest of Europe is based on the set of applications for healthcare environments grouped under the denomination of Soarian™ MedSuite [18, 29]. This circumstance limits its capacity of improvement.

3.3 Hewlett Packard: HCIS

The EHR architecture of HP is HP-HCIS. It is based on the conceptual model of HL7 V3 RIM, and it incorporates the typical elements of an EHR: clinical process, problems, documents, notes, orders, test results, medical images, multimedia, structured information of protocolized procedures, professionals visits, hospital stays, emergency services or external consultation, details of any type of attention, and hospital treatments or surgical operations.

This conceptual model, along with the use of controlled vocabularies (e.g. SNOMED), provides the EHR of HP a good chance to become a useful corporative work tool for the clinical daily work of healthcare professionals. This EHR can be made compatible with the inherited textual Health Record and facilitate the patient assistance, as well as offer a 24x7 access. It supports the clinical management and planning, and it serves as a basis for an installation of clinical guidelines or other elements of support to the installation of Evidence-Based Medicine, or to study and to evaluate the clinical variability.

Like SELENE, HP-HCIS is a solution that is only obsaerved in Spain, since in the rest of Europe HP does not have a specific proposal for Information Systems in healthcare environments. HP-HCIS is a solution that arises as an evolution of the applications developed during the 80s, in order to cover the need of Hospital Information Systems (HIS). It does not contemplate too many innovation elements.

3.4 CERNER

The EHR of CERNER Millenium [17] offers the more exhaustive range of functions of the sector. It is based on an expandable, unified and patient-centered architecture that allows a longitudinal EHR for whole institutions and multiple centers that include both functions and services.

All the authorized people (i.e., doctors, nurses, pharmacists, laboratory technicians and administrative staff) are able to access simultaneously to the same clinical record of a patient, an impossible task in a paper-based documentation system. Through a unique and integrated EHR, the information is always available where it is needed (see figure 5).

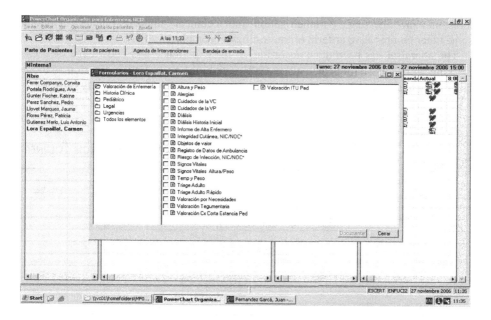

Fig. 5. CERNER Workstation

The main characteristics of CERNER are the combination of financial, operative and clinical information to carry out exhaustive analyses; the possibility of connecting distant centers and facilities, including intensive care units, semi-intensive, outpatient and rehabilitation units, and the access to an up-to-date and complete information of the patient in the way that better suits its needs.

CERNER Millenium is bursting with force into the Spanish healthcare market. Nevertheless, unlike the previous ones it does not have any great EHR project in that country, although there are several antecedents of use in healthcare centers in the United States.

Of all the previously described solutions, this is the more innovative because it incorporates Evidence-Based Medicine and Medical Decision Support elements. Along with IS-H*Med they could be classified as Computer-Based Patient Records (CPR) [13, 14] of the fourth generation (see table 1). This means that the EHR model proposed by these two tools is the most advanced of those exposed.

4   Comparative of the Considered Tools

When comparing the tools introduced in section 3, four types of comparative criteria have been established: functionality, use, installation and others. The criteria on the functionality of the systems include several desirable services that a knowledge management system in health should have (e.g., aid to the practice of evidence-based medicine). The criteria on the use include aspects that determine the application boundaries of the tools (e.g., degree of support for the different healthcare levels). The criteria on installation refer to several characteristics and experiences in putting the tools into practice (e.g., parametrization and technical support). Finally, in others some other comparative criteria of interest have been placed.

The appraisal has been done according to the experience and knowledge of IECI, marking with + those tools that successfully satisfy the comparative criterion, − those other tools that do not satisfy a criterion, and +/− if the criterion is partially satisfied.

Table 2. Comparative of the considered tools

Comparative criterion	SAP	SELENE	HCIS	CERNER
Feature				
Facilities for the accomplishment of clinical tests	−	+/−	−	+
It contains elements for Decision Support	+	−	−	+
It contains Data Mining tools	+/−	−	−	+
Connectivity with laboratory data	+	+	+/−	+
Connectivity with radiology data	+	+	+/−	+
Internet	+	+/−	−	+
It allows team work	+	+	+	+
Aid to healthcare processes control (workflows, protocols, etc.)	+	+	−	+
Interoperability with other systems	+	+/−	−	+
Doctor support	+	+	−	+
Nursing support	+/−	+	+	+
Manager support	+	−	−	−
Pharmacy support	+	−	−	+
Laboratory support	+	−	−	−
Facilities for explicit knowledge management	+	−	−	+
Facilities for implicit knowledge management	+	−	−	+
Facilities for tacit knowledge management	+/−	−	−	+/−
Facilities for Evidence-Based Medicine	+/−	+/−	−	+
Use				
Coverage of healthcare levels (hospitals, primary care, etc.)	+	+	+/−	−
Structural coverage (doctors, management, etc.)	+	+	+/−	−
Coverage of medical services (general med and specialties)	+/−	+	+/−	+
Professional coverage (doctor, nurse, etc.)	+/−	+	+/−	+
Adaptability to users	+	+/−	+/−	+
Adaptability to healthcare center	+	+/−	+/−	−
Adaptability to the national health system	+	+	+	−
Simplicity of use	+/−	+	+	−
Installation				
World-wide penetration degree	+/−	−	−	+
Europe penetration degree	+	−	−	+/−
Spain penetration degree	+/−	+/−	+/−	−
Installation simplicity	+/−	+	+/−	−
Good technical support	+	+	+/−	−
Cost proportional to product quality	+	+	+	−
Change management well supported	+/−	+	+	−
Other				
Security and reliability issues	+	+	−	+
Based on medical standards (codification, etc.)	+	+	+	+
Expectations of future of the tool	+	−	−	+

5 Experiences with the Previous Tools in the Spanish Health System

Since 2002 until the present moment, the execution of projects using these tools has had an unequal result in each Spanish Autonomous Region.

During the initial period between 2002 and 2004 the majority of health services in Spain started EHR project initiatives. During this period it was more important to have the tool than the way in which it was going to be used. Different healthcare systems in Spain adopted different solutions. As a result of this, during the period between 2004 and 2006, many of these projects had to be redefined (see figure 6, with the dark zones indicating different level of development of EHRs −only two of them fully installed, and clear zones as the regions without EHR projects).

Fig. 6. Map of the EHR installation in Spain in 2006

Other many regions were completing their political context or in full development of their EHR. The most adopted solution was SELENE, present in Murcia, Asturias, La Rioja and Madrid. On the other hand SAP IS-H*MED was the solution adopted by the Service of Health from Extremadura (SES) and the Catalan Health System (ICS). HP-HCIS was present in the Balearic Islands Health Service (IB-Salut) although only in the hospitals of Son Llatzer, Inca and Menorca. We should add that HP-HCIS was adopted by the national military health.

Fig. 7. Map of the EHR installation in Spain in 2008

At the end of 2008, the configuration of EHR in the Spanish territory is formed as it is indicated in figure 7, with eight regions with the EHR installed and six regions in process of development of the EHR (all of them dark zones).

4 Conclusions

Therefore considering that Health Records are the way to exchange knowledge and information between healthcare professionals at the point of care, the Electronic Health Record (EHR) as a tool can be viewed as a subsystem for healthcare knowledge management. There is a global consensus among all the implied healthcare actors, directors, professionals and users on the clear benefits of using EHR to face the management and treatment of diseases in modern Healthcare Systems. This fact has caused that the majority of the Spanish health services has prioritized the intallation of tools for the management of the EHR.

The technological tools on which these projects have been sustained not always have responded to the expectations and in many cases they have not been accompanied by the design of a project adapted to the real needs.

In Spain, some of the tools used in the health services of some Autonomous Regions are obsolete or they are not expected to have a high viability in the next future. However, considering the importance that EHR is acquiring as a compelling tool to reach a healthcare of quality, it is possible to anticipate that in the next five years all the European health systems will have tools to manage EHR. On the other hand, the compatibility between the different EHR is (and it will be) an important topic of research. In this context, we can expect that several additional functionalities will be added to the EHR of the future: medical decision and diagnostic support, detectors of possible incompatibilities in the treatment of patients, etc. [30].

Of the four systems for the EHR management, SELENE and HCIS are the ones that have more limitations at a general level. As far as the functionality is concerned, SAP and CERNER are the systems with a broader functionality. Concerning the use, all the systems are equivalent, unlike CERNER, clearly at a disadvantage. Concerning the installation, SAP surpasses the other systems.

References

1. García-Rojo, M.: Gestión del Conocimiento y las Nuevas Tecnologías de la Información en Salud. Complejo Hospitalario de Ciudad Real
2. Bravo, R.: La gestión del conocimiento en Medicina: a la búsqueda de la información perdida. Centro de Salud Sector III. Área 10 de Atención Primaria. Madrid
3. Ortú-Rubio, V.: Conocimiento para gestionar? Departamento de Economía y Empresa, Centro de Investigación en Economía y Salud
4. Drucker, P.E.: The information executives truly need. Harvard Business Review 1, 54–62 (1995)
5. Clinical Evicence. En: http://clinicalevidence.org
6. Ruiz, J., Meroño, A., Sabater, R.: Learning in organizations and information technologies. Their impact on organizational performance in small businesses. In: 4th European Conference on Knowledge Management (ECKM), MCIL, Oxford (2003)

7. Martínez-García, J.M.: Gestión intelectual en la empresa sanitaria. Publicado en Medical Economics. 3(11) (Junio de 2006)
8. Sánchez-Calas, J.C.: El informacionista clínico en el ámbito biomédico. Serie Bibliotecología y Gestión de Información 15, 1–54 (2006),
   http://eprints.rclis.org/6783/
9. SAP IS-H/IS-H*MED Release 4.71. SAP Library (April 2003),
   http://help.sap.com/saphelp_ish471/helpdata/EN/2e/
   109e377af82a17e10000009b38f842/frameset.htm
10. SAP for Healthcare. Healthcare Providers Industry Overview. White Paper (2007),
   http://www.sap.com/industries/healthcare/pdf/
   BWP_OV_SAP_for_Healthcare.pdf
11. Tapscott, D.: Business Intelligence for the Health Care Industry. Actionable Insights for Business Decision Makers (2008),
   http://www.sap.com/industries/healthcare/brochures/index.epx
12. Achaerandio-García, R.: SAP for Health care Solution Enables Hospital Clínico de Barcelona to Have a Consistent Overview. IDC Analyze the Future (2007)
13. Dick, R.S., Steen, E.B., Detmer, D.E. (eds.): The Computer-Based Patient Record: An Essential Technology for Health Care, Revised Edition (1997)
14. Erstad, T.L.: Analyzing Computer based Patient Records: A Review of Literature. Journal of Healthcare Information Management 17(4), 51–57 (2003)
15. https://www.swe.siemens.com/spain/internet/webs/areas/
   medical/oferta/Pages/selene.aspx
16. http://www.epiprensa.com/calidad-de-vida/
   hp-soluciones-informaticas.html
17. http://www.cerner.com/public/
18. http://www.medical.siemens.com/webapp/wcs/stores/servlet/
   CategoryDisplay~q_catalogId ~e_-18~a_categoryId~e_
   1008110~ a_catTree~e_100010,1008631,1007111,1007092,
   1008110~a_langId~e_-18~a_storeId~e_10001.htm
19. Dimond, B.: Electronic health record and electronic patient record. British Journal of Nursing 14(13), 716–717 (2005)
20. Casas-Valdes, A., Oramas Díaz, J., Presno Quesada, I., López Espinosa, J.A., Cañedo Andalia, R.: Theoretical aspects of knowledge management in evidence-based medicine. ACIMED 17(2) (2008)
21. Hopkins Tanne, J.: US government announces electronic health records project. BMJ 335(7627), 959 (2007)
22. The epSOS project, http://www.epsos.eu
23. Wena, H.-C., Hob, Y.-S., Jiana, W.-S., Lia, H.-C., Hsua, Y.-H.E.: Scientific production of electronic health record research, 1991–2005. Computer Methods and Programs in Biomedicine 86(2), 191–196 (2007)
24. Poissant, L., Pereira, J., Tamblyn, R., Kawasumi, Y.: Impact of Electronic Health Records on Time Efficiency of Physicians and Nurses: A Systematic Review. JAMIA 12(5) (2005)
25. Ambinder, E.P.: Electronic Health Records. Journal of Oncology Practice 1(2), 57–63 (2005)
26. Lobach, D.F., Detmer, D.E.: Research Challenges for Electronic Health Records. Am. J. Prev. Med. 32(5), 104–111 (2007)
27. Simon, S.R., Kaushal, R., Cleary, P.D., Jenter, C.A., Volk, L.A., Orav, E.J., Burdick, E., et al.: Physicians and Electronic Health Records. Arch. Intern. Med. 167, 507–512 (2007)

28. Implantación de un sistema de información de Historia Clínica Electrónica. Siemens
    Selene,
    `http://www.scribd.com/doc/6609110/`
    `Historiaclinicaelectronica-PDF-1465814`
29. `http://www.medical.siemens.com/webapp/wcs/stores/servlet/`
    `CategoryDisplay~q_catalogId ~e_-18~a_categoryId~e_`
    `1008109~a_catTree~e_100010,1008631,1007111,1007092,`
    `100810 9~a_langId~e_-18~a_storeId~e_10001.htm`
30. Sittig, D.G., Wight, A., Osheroff, J.A., et al.: Grand challenges in clinical decision support.
    J. of BioMed. Infor. 41, 387–392 (2008)

Author Index